# JTG Standards and Specifications

This series includes the standards and specifications of the highway transportation industry, issued by the Ministry of Transport of the People's Republic of China. It covers the standards and specifications in terms of technology, administration and service for the process from highway planning through to highway maintenance. The criteria for safety, environment and economic assessment are also included.

Hongyuan Wang

# Specifications for Highway Safety Audit

人民交通出版社
China Communications Press

Springer

Hongyuan Wang
Chelbi Engineering Consultants Inc.
Beijing, China

ISSN 3004-8818 ISSN 3004-8826 (electronic)
JTG Standards and Specifications
ISBN 978-981-92-3131-7 ISBN 978-981-92-3132-4 (eBook)
https://doi.org/10.1007/978-981-92-3132-4

Jointly published with China Communications Press Co., Ltd
The print edition is not for sale in China (Mainland). Customers from China (Mainland) please order the print book from: China Communications Press Co., Ltd



This Springer imprint is published by the registered company Springer Nature Singapore Pte Ltd.
The registered company address is: 152 Beach Road, #21-01/04 Gateway East, Singapore 189721, Singapore

# Foreword to Chinese Version

In order to comply with the requirement of *Notice on Plans for the Formulation and Revision of Highway Engineering Standards* in 2011 (Gong Lu Zi [2011] No. 115) issued by the Ministry of Transport of the People's Republic of China, Chelbi Engineering Consultants Inc. was appointed to formulate and revise the *Guidelines for Safety Audit of Highway* (JTG/T B05-2004) (herein after referred to as 'the *Guidelines*').

The *Guidelines* has served as the first recommended standard of safety audit in the transportation industry. The *Guidelines* has been important for improving highway facilities and traffic safety conditions and promoting improved highway safety since its implementation in November 2004. With the rapid development of the highway transportation industry and a growing demand for traffic safety in China, strengthening and promoting the application of highway safety audit has become inevitable.

Based on recent research and practical experience in China and foreign countries, development of the current focus safety audit, and the requirement to develop a framework for 'Safe Transportation', the *Guidelines* has been comprehensively revised and amended. *Specifications for Highway Safety Audit* (hereinafter referred to as 'the *Specifications*') (JTG B05-2015) is issued for implementation accordingly.

The *Specifications* consists of seven chapters and three appendixes.

1. General Provisions;
2. Terms and Definitions;
3. Feasibility Stage;
4. Preliminary Design Stage;
5. Detailed Design Stage;
6. Pre-opening Stage;
7. Operation Stage;

Appendix A: Format of Safety Audit Report;
Appendix B: Prediction Methods for Operating Speed;
Appendix C: Calculation Methods for Clear Zone Width.

The main amendments are shown as follows:

1. The focus of safety audit and requirements of audit process at each stage are expanded.
2. The requirements for contents and details of conclusion of safety audit at each stage are added.
3. Contents related to the audit of Class-2 highways, Class-3 highways, and highway upgrading and reconstruction projects at each stage are added.
4. Adjustment has been made to the content of chapters related to design stage, which have been compiled according to the stages of Preliminary Design and Detailed Design.
5. The chapter of the pre-opening stage is added.
6. Adjustment has been made to the partial contents of the chapter of Operation Stage Audit, while the contents of post-audit are incorporated into the chapter of Operation Stage.
7. The methods of predicting operating speeds of motorways, Class-1 highways, Class-2 highways and Class-3 highways and are appended and improved.
8. Format of Safety Audit Report is adjusted.

As possible input for the next revision, comments and suggestions from users during the process of implementation are welcome and should be addressed in time to Chelbi Engineering Consultants Inc. (Address: F8, Century Xingyuan Building, No. 20, Anyuan Road, Chaoyang District, Beijing; Postcode: 100029; Tel: 010-84898688; Fax: 010-84896981; E-mail: safety@huajie.com.cn).

Editing Organization in charge:

Chelbi Engineering Consultants Inc.

Associate Editing Organizations:

Research Institute of Highway Ministry of Transport

Tongji University

Chief Editor:

Wang Hongyuan

Associate Editors:

Zhong Xiaoming, Zhou Ronggui, Jia Jia, Fang Jing, Tang Chengcheng, Guo Zhongyin, Yang Zhen

Chief Reviewer:

Chen Yongyao

Associate Reviewers:

Bao Gang, Chen Biao, Chen Jianzhuang, Guo Min, Guo Tengfeng, Han Fengchun, Huang Yuetong, Jiang Yousheng, Liao Chaohua, Lin Fei, Liu Bing, Liu Huixue, Rong Jian, Sun Xiaoduan, Tu Yun, Wang Jing, Yang Chunhui, Zhan Hui, Zhang Meichai, Zhuang Lingyun

Beijing, China

# Introduction to English Version

Standards reflect the achievement of civilization, provide common language for technical communications, and improve global connectivity. In recent years, the Chinese government has been proactively implementing a strategy on standardization to stimulate innovation, coordination, greening, opening up and sharing for reciprocal development in China and worldwide. In the light of mutual development along the Silk Road Economic Belt and the 21st-Century Maritime Silk Road (so called 'the Belt and Road Initiative'), the Ministry of Transport of the People's Republic of China organized translation and published an international version of the Chinese transportation industry standards and specifications to cater for the increasing demands for international cooperation in world transportation, achieve interconnected development and promote knowledge dispersion and sharing experience.

JTG is the designation referring to the standards and specifications of the highway transportation industry, issued by the Ministry of Transport of the People's Republic of China. It covers the standards and specifications in terms of technology, administration and service for the process from highway planning through to highway maintenance. The criteria for safety, environment and economic assessment are also included.

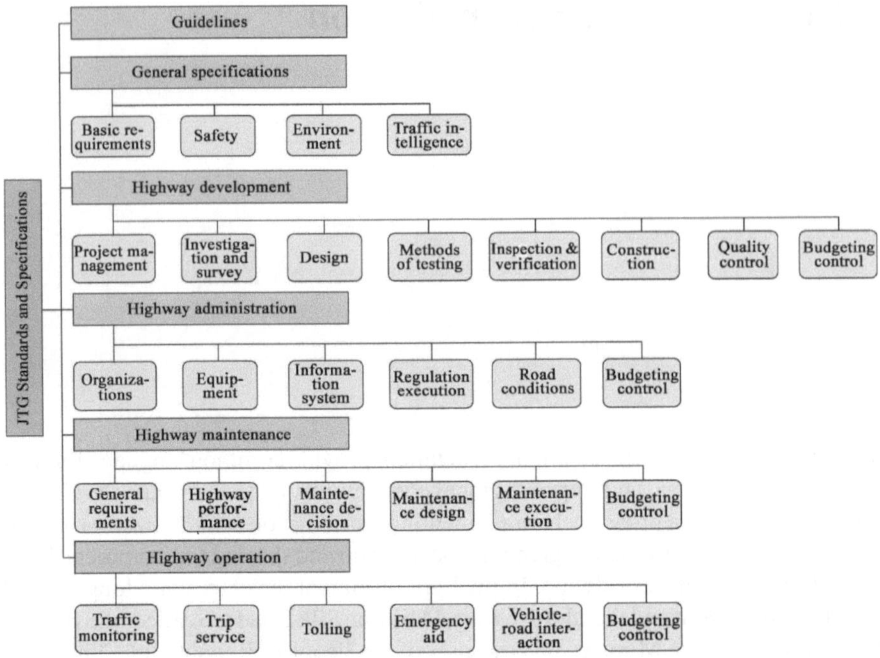


The Chinese government has always attached great importance to road traffic safety issues. As one of the important participating countries of the United Nations 'Decade of Action for Road Safety 2011–2020', it has deployed and implemented a series of targeted measures, including improving relevant laws, policies, standards and specifications, implementation of 'Highway Safety Protection Project' and 'Improve Highway Safety to Cherish the Life Projects', etc., promoting the application of 'Highway Safety Audit Projects'. As engineer experience and scientific research achievements in China were involved in the standards and specifications, highway safety has promoted in highway design, construction and maintenance. The number of traffic crashes has greatly decreased, and traffic safety situation has become favorable for road users in recent years.

The Ministry of Transport (MOT) of the People's Republic of China issued the *Guidelines for Safety Audit of Highway* (JTG/T B05-2004) in 2004. It was revised and updated as the *Specifications for Highway Safety Audit* (JTG B05-2015) in 2015. This upgrade version has well absorbed China's scientific research achievements and Road Safety Audit (RSA) experience over the past 20 years, and has referred international advanced technologies and standards and regulations. And the concept of RSA has been substantially accepted and spread by the authorities and engineers in China. Highway safety performance has enhanced observably after abroad application of RSA. RSA plays critical roles in development of highway facilities, improving safe operations, reducing the risk of crashes, and promoting road safety. The RSA projects have been implemented on more than 50,000 km of highway in each year in China.

The English version of *Specifications for Highway Safety Audit* (JTG B05—2015) was conducted by Chelbi Engineering Consultants Inc. authorized by the MOT and approved by Highway Department of the MOT. Several international experts assisted during the editing and approving of the English version. Gratitude is given here to the British expert John Barrell and the American expert W. Gregory Wood for the final review and proofreading of the English version. Meanwhile, thanks also given to the editors Guo Zhongyin, Yang Zhen, and the reviewer Chen Yongyao in the Chinese version.

Comments, suggestions and inquiries are welcome and should be addressed to Chelbi Engineering Consultants Inc. (Address: F8, Century Xingyuan Building, No. 20, Anyuan Road, Chaoyang District, Beijing; Postal code: 100029; Tel: 010-84898688; Fax: 010-84896981; E-mail: safety@huajie.com.cn). The feedback shall be taken into account in future editions.

Editing Organization for English version: Chelbi Engineering Consultants Inc.

Chief Editor for English version:

Wang Hongyuan

Associate Editors for English version:

Zhong Xiaoming, Tang Chengcheng, Jia Jia, Li Ruimei, Zhou Juan, Zhang Jian, Yu Hao, Zhang Xiaoqin, Lu Tong, Fang Jing, Zhou Jian, Li Wei, Zhou Ronggui

Reviewers for English version:

Wu Dejin, Zhang Jianjun, Zhang Huiyu, Xie Shaozhang, Yao Haidong, Kwong Tse Hin Glenn, Si Xiaowei, Chai Hua, Zhang Tiejun, Hu Han, John Barrell, W. Gregory Wood

# The People's Republic of China, Ministry of Transport: Public Notice

***No. 61***

**Public Notice on Issuing the English Version of *Specifications for Highway Safety Audit***

The English version of *Specifications for Highway Safety Audit* [JTG B05-2015 (EN)] is hereby issued for international cooperation and standardization of highway transportation industry.

The general administration and final interpretation of the English version of the *Specifications* belong to Ministry of Transport of the People's Republic of China, while particular interpretation for application and routine administration of the English version of the *Specifications* shall be provided by Chelbi Engineering Consultants Inc.

In the event of any ambiguity or discrepancies between the English version and the Chinese version of the *Specifications*, the Chinese version shall be referred to and accepted.

Comments, suggestions and inquiries are welcome and should be addressed to Chelbi Engineering Consultants Inc. (Address: F8, Century Xingyuan Building, No. 20, Anyuan Road, Chaoyang District, Beijing; Postal Code: 100029; Tel: 010-84898688; Fax: 010-84896981; E-mail: safety@huajie.com.cn). The feedbacks will be considered in future editions.

It is hereby announced.

Ministry of Transport of the People's Republic of China

July 31, 2020

# The People's Republic of China, Ministry of Transport: Public Notice

***No. 59***

**Public Notice on Issuing the *Specifications for Highway Safety Audit***

Hereby the *Specifications for Highway Safety Audit* (JTG B05-2015) is issued as one of the highway engineering industry standards, to become effective from April 1, 2016. The former edition of the *Guidelines for Safety Audit of Highway* (JTG/T B05-2004) shall be superseded from the same date.

The Ministry of Transport holds the right to manage and interpret the *Specifications for Highway Safety Audit* (JTG B05-2015), while Chelbi Engineering Consultants Inc. is responsible for routine interpretation and management.

Comments and suggestions from users are welcome and should be addressed to Chelbi Engineering Consultants Inc. (Address: F8, Century Xingyuan Building, No. 20, Anyuan Road, Chaoyang District, Beijing; Postal Code: 100029). The feedbacks will be considered in future editions.

It is hereby announced.

Ministry of Transport of the People's Republic of China
December 23, 2015

# Contents

# Chapter 1
# General Provisions

1.0.1 The *Specifications* is formulated to standardize and regulate highway safety audit.

1.0.2 The *Specifications* is applicable to the implementation of safety audit of motorways and Class-1, Class-2 and Class-3 highways.

1.0.3 The *Specifications* is applicable to the stages of feasibility study, preliminary design, detailed design, pre-opening and operation of highways.

1.0.4 The representative vehicle types used for the purpose of safety audit shall have designs stipulated in the *Technical Standards for Highway Engineering* (JTG B01-2014), taking into account the actual traffic composition on the highway.

1.0.5 In addition to the *Specifications*, highway safety audit shall also comply with the relevant current national and industry standards.



H. Wang, *Specifications for Highway Safety Audit*, JTG Standards and Specifications,
https://doi.org/10.1007/978-981-92-3132-4_1

# Chapter 2
# Terms and Definitions

2.0.1 Road safety audit, or highway safety audit

This is the comprehensive and systematic analysis and audit of highway traffic safety from the perspective of all highway users following a certain audit procedure and using qualitative and quantitative methods, hereinafter referred as RSA.

2.0.2 Operating speed-based design consistency

An indicator to evaluate the design consistency of alignment geometry, using operating speed differentials between successive geometric feature segments, as well as the speed differentials between operating speed and design speed on the same segments.

2.0.3 Safety audit checklist

A checklist of safety audit items that need to be considered at each stage of audit, prepared according to crash analysis of prevention, design standards and engineering experience of highway safety, etc.



H. Wang, *Specifications for Highway Safety Audit*, JTG Standards and Specifications,
https://doi.org/10.1007/978-981-92-3132-4_2

# Chapter 3
# Feasibility Stage

## 3.1 General

3.1.1 The focus of audit at this stage shall be the traffic safety implications of corridors and engineering proposals on road users, society and the environment.

3.1.2 Newly constructed highways shall be assessed with respect to alternative corridor proposals with the same level of details.

3.1.3 For highway upgrading and reconstruction projects, the characteristics of traffic safety of the existing highway shall be analyzed, and the implications of any alternative schemes of highway upgrading on traffic safety shall be assessed.

## 3.2 Methodology

3.2.1 This stage should be assessed by empirical analyses or using safety audit checklists.

3.2.2 The analysis of the traffic safety characteristics of existing highways for highway upgrading and reconstruction projects shall comply with the relevant requirements in Sections 7.3 and 7.4 of the *Specifications*.

## 3.3 Audit Scope

3.3.1 The audit of engineering proposals shall be in compliance with the requirements as follows:



H. Wang, *Specifications for Highway Safety Audit*, JTG Standards and Specifications,
https://doi.org/10.1007/978-981-92-3132-4_3

1. The implications of the engineering construction on traffic safety shall be assessed based on terrain conditions and traffic composition, etc. The implications on traffic safety shall be assessed with respect to upgraded and reconstructed highway.
2. The implications of the form of connection and the traffic management plan, etc. between origin and destination with other highways on traffic safety shall be assessed according to the traffic volume.
3. The implications of hazardous segments with sharp curves and steep grades, continuous upgrades, continuous long and steep descending, road segments adjacent to cliffs, deep ravines and water bodies etc. on traffic safety shall be assessed.
4. The implications of the site selection, scale and operational safety requirements of super-large bridges, extra-long tunnels and other large structures on traffic safety shall be assessed.
5. The implications of site selection and form of interchanges, as well as the interaction between adjacent interchanges, between interchanges and large structures such as tunnels and between administration and service facilities on traffic safety shall be assessed according to the conditions of road network, traffic volume, composition and the layout of cities and towns along the routes, etc.
6. The implications of site selection, forms, traffic management plan and spacing of at-grade intersections shall be assessed according to terrain conditions, technical criteria of the main routes, conditions of the intersected highways and forecast traffic volumes, etc.
7. The implications of railways, oil and gas pipelines and high-voltage transmission lines crossing or passing in the vicinity of the project on traffic safety shall be assessed.
8. The implications of roadside interference on traffic safety shall be assessed based on the conditions of passage through villages and towns, residential areas, pasture and forest areas, etc.
9. The traffic management proposals of highway upgrading and reconstruction projects and the corresponding safety measures shall be assessed when traffic flow is maintained during the construction period or the mainline traffic is diverted to connecting roads.

3.3.2 The implications of environmental conditions on traffic safety shall be assessed based on climate conditions such as rainfall, freezing, snow, fog and crosswind.

3.3.3 The ability of the proposed route and connecting roads to provide adequate Evacuation Emergency and Rescue (EER) facilities shall be assessed when traffic disruption occurs in the event of natural disasters or serious traffic accidents.

3.3.4 The necessity to provide fences or passages for animals shall be assessed based on habitats of animals and routes for animal migration.

## 3.4 Findings and Recommendations

3.4.1 In the audit conclusions, the results of safety analyses shall be listed, the key issues that affect the traffic safety of highways shall be clarified, and suggestions for appropriate countermeasures for the next stages of designs shall be recommended.

3.4.2 For highway upgrading and reconstruction projects, the audit conclusions shall address whether existing critical issues of traffic safety can be alleviated or resolved.

# Chapter 4
# Preliminary Design Stage

## 4.1 General

4.1.1 This audit shall focus on the route proposals and applications of technical standards, the rational provision and layout of structures (bridges, tunnels, interchanges, etc.) and the rationality of scale of construction of traffic control devices and roadside facilities.

4.1.2 Overall audit, alternatives audit and audit on design elements shall be carried out. Alternatives audit shall be conducted based on submitted proposals with the same level of details. Audit on design elements shall be conducted based on the recommended proposal.

4.1.3 The safety audit of preliminary design stage shall be conducted according to the *Specifications*.

The *Format of the Highway Safety Audit Report* shall conform to the relevant requirements in Appendix A of the *Specifications*.

## 4.2 Methodology

4.2.1 The alternatives audit should be carried out using empirical analysis or safety audit checklist.

4.2.2 The audit on design elements may be based on methods such as the analyses of operating speed-based design consistency on traffic safety.



H. Wang, *Specifications for Highway Safety Audit*, JTG Standards and Specifications,
https://doi.org/10.1007/978-981-92-3132-4_4

## 4.3 Overall Audit

4.3.1 The implications of highway project characteristics on traffic safety shall be audited according to relevant technical standards, terrain, geological and climatic conditions, forecast traffic volume and traffic composition as well as the distribution of large structures (bridges, tunnels, interchanges, etc.).

4.3.2 If existing road sections are reused in highway upgrading and reconstruction projects, the characteristics of such road sections shall be analyzed and the design criteria of the reused road sections shall be audited by the current technical standards based on the operational conditions of existing highways and accident data.

4.3.3 The implementation of comments related to traffic safety shall be checked in the approval of engineering feasibility study.

4.3.4 If an audit has been carried out at the feasibility stage, the actions in response to the comments of the safety audit shall be checked.

## 4.4 Alternatives Audit

4.4.1 The implications of segments with sharp curves and steep grades, continuous upgrades, continuous long downgrades, road segments adjacent to cliffs, deep ravines and water bodies as well as other dangerous segments on traffic safety shall be assessed for each alternative proposal.

4.4.2 The implications of overall distribution and consistency of super-large bridges, extra-long tunnels, group of tunnels, interchanges, major intersections, as well as service facilities, etc. on traffic safety shall be assessed for each alternative proposal.

4.4.3 The implications of adverse weather or environment on traffic safety shall be considered for each proposal.

4.4.4 For highway upgrading and reconstruction projects, the implications on traffic safety shall be assessed with respect to the continuity of alignment and treatments at roadway widening, and the reuse of existing traffic safety devices for each alternative.

## 4.5 Audit on Design Elements

4.5.1 The operating speed-based design consistency shall be evaluated for highways with design speed at or below 80 km/h. The evaluation shall be in compliance with the requirements as follows:

1. The evaluation of operating speed-based design consistency shall include evaluating operating speed consistency (hereinafter referred to as OSC) between successive geometric feature segments, and the speed consistency (hereinafter referred to as SC) between operating speed and design speed on the same segment.
2. Operating speeds shall be predicted with the method provided in Appendix B of the *Specification.* The parameter calibration of the calculation model shall be carried out according to the characteristics of the location of the projects. If the conditions are not available, the relevant parameters may be taken according to Appendix B of the *Specifications.*
3. OSC shall be evaluated by an absolute value of operating speed differential in successive segments $|\Delta v_{85}|$, and an absolute value of operating speed gradient $|\Delta I_v|$. The criteria of design consistency evaluation at successive geometric feature segments shall comply with the requirements in Table 4.5.1.
4. SC shall be evaluated by the speed differentials between operating speed and design speed on the same segments. If the difference is greater than 20 km/h, the technical criteria of the road sections shall be evaluated according to the operating speeds.
5. For highway upgrading and reconstruction projects, operating speed-based design consistency shall be integrally evaluated for the new road sections and the existing road sections.

4.5.2 Alignment shall be audited in compliance with the requirements as follows:

1. The evaluation of highway layout plan shall be in compliance with the requirements as follows:
   (1) Circular curves close to the minimum radius shall be evaluated according to the operating speeds.
   (2) The parameters and length of spiral curve, as well as the length of tangent sections between curves and the length of horizontal curves should be evaluated based on operating speeds and visual conditions etc.
   (3) The continuity and balance of the alignment upstream and downstream of U curves and the distance between reverse curves (switchback) shall be evaluated.
   (4) Special curves such as oval curves and compound curves etc. should be evaluated.

2. Sight distance shall be evaluated in compliance with the requirements as follows:
   (1) The stopping sight distance shall be evaluated for motorways and Class-1 highways; the meeting sight distance, and the passing sight distance shall be evaluated for Class-2 highways and Class-3 highways.
   (2) Relevant road sections of motorways and Class-1 highways as well as Class-2 and Class-3 highways with a higher proportion of large vehicles shall also be evaluated with the stopping sight distance of trucks.
   (3) Stopping sight distance, meeting sight distance and overtaking sight distance should be evaluated using operating speeds.

**Table 4.5.1** Evaluation criteria of operating speed-based design consistency at adjacent segments

| Operating speed-based design consistency at adjacent segments | Evaluation criterion | Solutions and suggestions |
|---|---|---|
| Motorways and Class-1 highways | | |
| Good | $\|\Delta v_{85}\| < 10$ km/h and $\|\Delta I_v\| \leq 10$ km/(h·m) | |
| Fair | 10 km/h $\leq \|\Delta v_{85}\| < 20$ km/h and $\|\Delta I_v\| \leq 10$ km/(h·m) | If decelerating at successive road sections, the horizontal and vertical profile design of successive road sections should be optimized or safety improvements shall be taken |
| Poor | $\|\Delta v_{85}\| \geq 20$ km/h or $\|\Delta I_v\| > 10$ km/(h·m) | If decelerating at adjacent sections, the horizontal and vertical profile design of successive road sections shall be adjusted; If it is difficult to make adjustments, safety improvements shall be taken |
| Class-2 highways and Class-3 highways | | |
| Good | $\|\Delta v_{85}\| < 20$ km/h and$\|\Delta I_v\| \leq 15$ km/(h·m) | |
| Poor | $\|\Delta v_{85}\| \geq 20$ km/h or $\|\Delta I_v\| > 15$ km/h·m | If decelerating at adjacent segments, the horizontal and vertical alignment and profile design of successive road sections shall be adjusted, or safety improvements shall be taken |

3. The highway profile shall be evaluated in compliance with the requirements as follows:

   (1) Continuous upgrades and downgrades shall be evaluated.
   (2) Vertical curves close to minimum radius or minimum length should be evaluated according to the operating speeds.

4. Highway cross sections shall be evaluated in compliance with the requirements as follows:

   (1) The location and length of transition section shall be evaluated if there is a change in width of cross sections or number of lanes.
   (2) For continuous upgrades, the necessity of providing climbing lanes and their locations shall be considered according to forecast traffic volume and traffic composition, level of service and operating speeds etc.
   (3) For continuously long steep downgrades, the necessity of providing emergency escape ramps, their locations and quantities shall be considered, based

on forecast traffic volume and traffic composition, terrain conditions and locations of service facilities, etc.

(4) If the width of hard shoulder on the right side of motorways and Class-1 highways is less than 2.5 m, the effective length, width and spacing between any adjacent emergency lay-bys, and their entrance/exit transition sections shall be evaluated.

(5) For road sections with high numbers of non-motorized vehicles and pedestrians, the conditions of roadside interference and provision of non-motorized vehicle lanes and sidewalks should be considered.

(6) For highways or entrance/exit to cities with high density of non-motorized vehicles and pedestrians, the implications of mixed traffic on traffic safety should be assessed.

5. For highway upgrading and reconstruction projects, the locations of diverges and merges of the mainline and the balance in the number of lanes shall be evaluated.

4.5.3 Roadside shall be audited in compliance with the requirements as follows:

1. The width of roadside clear zones and the roadside risk level shall be evaluated according to the operating speeds. The Clear Zone Width may be determined by the method provided in Appendix C of the *Specifications*.
2. The necessity of adopting measures such as roadside protection, removal or relocation of roadside obstacles shall be assessed.

4.5.4 Bridges shall be audited in compliance with the requirements as follows:

1. The implications of geometric alignment design of bridge approaches and bridge sections on traffic safety shall be evaluated in conjunction with the conditions at the bridge location.
2. If the cross-sectional width of the bridge approach is different from that of the bridge, the provision of transition sections and their length shall be evaluated.
3. If a long and large bridge is not provided with a hard shoulder, the necessity of providing emergency lay-bys shall be considered according to the needs for traffic safety.
4. For bridges crossing over the audited highway project, the implications of bridge piers, abutment and the superstructure on sight distance shall be evaluated.

4.5.5 Tunnels shall be audited in compliance with the requirements as follows:

1. The design consistency of geometric alignments inside and outside the tunnel portals should be evaluated according to the operating speed.
2. If the tunnel portals are located on vertical curves, the implications on drainage shall be considered.
3. The connection and transition between the cross-sections of the tunnel and tunnel approaches shall be assessed.
4. The sight distance of tunnels on curves shall be evaluated according to the operating speed.

5. The implications of the orientation and form of tunnel portals on traffic safety shall be considered.
6. The evaluation of upgrading and reconstruction projects of existing tunnels shall also be in compliance with the requirements as follows:

   (1) If existing highway tunnels are retained, the statistical data of traffic crashes shall be used to analyze the cause of crashes and their relationship with the tunnel alignment, civil works, traffic control devices and ancillary facilities.
   (2) If the design speed is increased, the implications of the clearance profile of existing highway tunnels on traffic safety shall be reviewed.

4.5.6 Interchanges shall be audited in compliance with the requirements as follows:

1. The location and form of interchanges shall be assessed according to the topographical conditions of intersecting highways, horizontal and vertical alignment criteria of intersected roads and turning traffic volume, etc.
2. The spacing between adjacent interchanges, between an interchange and adjacent service areas, tunnels and toll stations on main route, etc. shall be evaluated.
3. The form of interchange entrances or exits shall be assessed according to the classification of intersected highways, turning traffic volume, terrain conditions and the method of toll collection, etc.
4. If the speed differential between the operating and design speeds on the mainline is greater than 20 km/h, the sight distance of interchanges, the spacing between successive exits, the length of acceleration and deceleration lanes, etc. shall be evaluated according to the operating speeds.
5. Highway capacity and level of service, etc. may be evaluated according to the scale and traffic volumes at interchanges.
6. The evaluation of upgrading or reconstruction projects of existing highways shall comply with the following requirements:

   (1) If a new interchange is to be built, the spacing between the new interchange and other facilities or structures shall be evaluated.
   (2) For the upgrading or reconstruction of existing interchanges, the proposal shall be assessed in relation to forecast traffic volume and crash studies, etc.

4.5.7 Intersections shall be audited in compliance with the requirements as follows:

1. The position and spacings of adjacent intersections shall be assessed according to terrain conditions, horizontal and vertical alignment of the mainline, road network layout and conditions of intersecting highways, etc. The feasibility of combining closely spaced intersections shall be assessed.
2. The form of intersections shall be assessed based on turning traffic volume, classes of intersecting highways, modes of traffic management and distribution of adjacent roads, etc.
3. Adopted measures of speed control and traffic management shall be assessed according to the relevant operating speeds.

4. The visibility of sight triangles at intersections shall be evaluated in conjunction with the method of traffic control and operating speeds.

4.5.8 Traffic control devices and roadside facilities shall be audited in compliance with the requirements as follows:

1. The design principle and type of safety devices (including traffic signs, markings, barriers, visual guiding devices, and anti-glare devices, etc.) and their suitability for the overall project shall be assessed according to traffic volume and traffic composition, alignment, operating speeds and weather conditions, etc.
2. Service areas and rest areas shall be evaluated in compliance with the requirements as follows:

   (1) The locations and spacing of service areas and rest areas shall be evaluated according to the overall layout of highway service facilities, traffic volume and composition, important structures and continuous longitudinal grades, etc.

   (2) The scale of service areas and rest areas shall be evaluated according to the traffic volume, traffic composition as well as land use planning, etc.

   (3) The geometric alignment of on-ramps and off-ramps, sight distance and length of acceleration or deceleration lanes of service areas and rest areas shall be evaluated according to the operating speeds.

3. Toll stations shall be evaluated in compliance with the requirements as follows:

   (1) The location of toll stations shall be assessed according to terrain conditions, traffic volume and traffic composition, spacing between ramp toll stations and the nose of a gore area or an entrance ramp (or diverging and converging point) as well as spacing between intersections and spacing between mainline toll stations and tunnels, etc.

   (2) For toll stations located at the end of continuous long downgrades and ramps or after a sharp curve, the feasibility of relocation shall be assessed. If conditions are constrained, facilities of safety protection and speed management shall be considered.

   (3) The visibility of toll stations along the mainline and the ramps shall be evaluated in accordance with the operating speeds and the stopping sight distance of large vehicles.

4. The location, the sight distance, entrances and exits of checkpoints shall be evaluated.
5. The design principles, quantity and forms of surveillance and control facilities, etc. shall be assessed based on highway class, traffic volume and traffic composition, distribution of important structures (bridges and tunnels, etc.), and distribution of road sections susceptible to weather hazards and continuous long downgrades, taking into account requirements for internet and sustainable development.
6. The evaluation of highway upgrading or reconstruction projects shall be in compliance with the requirements as follows:

(1) If highway upgrading and reconstruction projects retain existing continuous long downgrades, road sections with constrained horizontal and vertical alignment features, sections of diverging and merging, and the sections susceptible to weather hazards, etc., then comprehensive improvement measures shall be assessed.
(2) If new service facilities are proposed, the spacing between the new service facilities and other facilities or structures and provision of safety devices shall be evaluated.

## 4.6 Findings and Recommendations

4.6.1 The audit conclusions shall include findings and recommendations for the overall audit, alternatives audit and audit on design elements.

4.6.2 The conclusions of overall audit shall illustrate the characteristics of the highways and their implications on traffic safety.

4.6.3 The conclusions of alternatives audit shall illustrate the safety results on alternative routes at the same level of detail, and shall recommend the preferred route proposal based on safety as a priority.

4.6.4 The conclusions of the audit on design elements shall present improvement recommendations and suggested countermeasures for each of the identified problems.

4.6.5 If the conclusions of the audit on design elements contain a number of improvement recommendations and suggested countermeasures, particularly involving changes of the design proposals, the sequence of implementation should be provided based on their level of implications on traffic safety.

# Chapter 5
# Detailed Design Stage

## 5.1 General

5.1.1 This audit shall focus on layout and the placement conditions of traffic control devices and roadside facilities.

5.1.2 Overall audit and audit on design elements shall be carried out.

5.1.3 For highway upgrading and reconstruction projects, the safety implications of the traffic management plan in the construction period shall be assessed.

5.1.4 For highway projects based on single stage detailed design or without any safety audit at the preliminary design stage, the audit on design elements shall be carried out in accordance with the relevant requirements in Section 4.5 of the *Specifications* and shall comply with the relevant requirements in this Chapter.

5.1.5 The safety audit for the detailed design stage of highway projects shall be conducted according to the *Specifications.* The format of the *Highway Safety Audit Report* shall conform to the relevant requirements in Appendix A of the *Specifications.*

## 5.2 Methodology

5.2.1 At this stage evaluation methods such as operating speed-based design consistency and safety audit checklists should be adopted.

5.2.2 For complex projects and road segments, driving simulation may be used to evaluate the design consistency of alignment and traffic safety devices, etc.



H. Wang, *Specifications for Highway Safety Audit*, JTG Standards and Specifications,
https://doi.org/10.1007/978-981-92-3132-4_5

## 5.3 Overall Audit

5.3.1 The characteristics of highway project on traffic safety shall be analyzed and this shall be in compliance with relevant requirements in Clause 4.3.1 of the *Specifications.*

5.3.2 The implementation of comments related to traffic safety should be checked in the approval of preliminary design.

5.3.3 If an audit has been carried out at the preliminary design stage, the actions in response to the comments of the safety audit shall be checked.

## 5.4 Audit on Design Elements

5.4.1 Alignment shall be audited in compliance with the requirements as follows:

1. Super-elevation design shall be evaluated in compliance with the requirements as follows:
   (1) For a given radius of circular curve, super-elevation shall be evaluated according to the operating speed.
   (2) The maximum super-elevation value shall be evaluated based on factors such as highway classes, regional climate conditions and traffic composition.
   (3) For highways with a high proportion of large vehicles, super-elevation values shall be evaluated in accordance with the speed differential between various vehicle types, and the implications of steep downgrades on the super-elevation values.
2. If the widening is applied to circular curves, widening values and the type of widening shall be evaluated according to traffic composition.
3. Combinations of super-elevation and longitudinal grades shall be evaluated with operating speeds according to climatic and terrain conditions as well as traffic composition.
4. For road sections designed with climbing lanes, the length and width of climbing lanes, their locations and the quantities of lanes, and the associated signs and road markings as well as other factors shall be evaluated.
5. For road sections designed with emergency escape ramps, the location, quantities and spacings of emergency escape ramps shall be evaluated. Moreover, the approach lanes, horizontal alignment and vertical alignment, cross-section width, length and gradient of ramps, materials of arrester bed, drainage, terminal treatments, as well as safety devices and ancillary facilities shall be considered.
6. Highway upgrading and reconstruction project shall be evaluated in compliance with the requirements as follows:

(1) For upgrading of existing highway sections with changes of driving direction, the rationality of reusing the existing super-elevation shall be analyzed and safety measures shall be evaluated according to the technical criteria for the actual alignment.
(2) If the highway is widened on one side, the location and length of transitions at lane change over zones and the associated traffic control devices shall be evaluated.

5.4.2 Subgrade and pavement shall be audited in compliance with requirements as follows:

1. For interface sections with different pavement materials, or sections prone to reduction in skid resistance, measures for increasing the pavement skid resistance should be assessed.
2. The location of median openings shall be evaluated.
3. The drainage facilities shall be evaluated in compliance with the requirements as follows:

   (1) If a side ditch or a gutter is located within the suggested clear zone, their forms shall be assessed.
   (2) In areas with heavy rainfall, the type of pavement drainage, drainage facilities at the bottom of sag vertical curves, sections on super-elevation and super-elevation transition section should be evaluated.

4. For highway upgrading and reconstruction projects, the function of existing drainage facilities and the improved design should also be assessed.

5.4.3 Bridges and culverts shall be audited in compliance with the requirements as follows:

1. For bridges crossing over the audited highway project, the implications of the location of piers and abutments on traffic safety shall be assessed according to their relationship with roadside clear zones.
2. Improvement of the skid resistance of bridge deck pavement should be assessed according to the operating speeds in conjunction with the longitudinal gradients and cross slopes of bridges.
3. The drainage capacity of discharge holes on bridge decks should be reviewed according to the intensity of rainfall and the longitudinal gradients of bridges, and the implications of water dropping from the bridge deck on vehicle traffic and pedestrians should be considered.
4. If a bridge is located in an area of prevailing high winds, the implications of crosswinds on traffic safety over the bridge deck shall be considered.
5. If median openings are provided in a super-large bridge for EER, the appropriateness of their locations shall be assessed.
6. If unshielded culvert openings are located within the suggested roadside clear zone and the roadside is not provided with any roadside barriers, the implications of the form of culvert openings on traffic safety shall be assessed.

5.4.4 Tunnels shall be audited in compliance with the requirements as follows:

1. The location, number and intersecting angle of cross passages for vehicles or pedestrians shall be assessed.
2. If there are differences in the skid resistance of pavements in and outside a tunnel, improvement measures for skid resistance at the tunnel portal should be assessed.
3. The implications of tunnel lighting, ventilation, fire-fighting equipment, surveillance and control facilities on traffic safety should be evaluated.
4. The evaluation of EER for tunnels shall be in compliance with the requirements as follows:

   (1) The implications of the positioning of crossovers and ancillary facilities at the portals of separate tunnel tubes on traffic safety should be assessed considering the alignment geometry and sight distance at the tunnel portals.
   (2) The conditions of EER of long tunnels, extra-long tunnels and chains of tunnels should be considered.

5. The evaluation of highway tunnel upgrading and reconstruction projects shall be in compliance with the requirements as follows:

   (1) Proposals of ventilation, lighting, traffic safety, surveillance and fire-fighting facilities, etc. should be assessed according to the capacity and safety conditions of upgraded tunnels.
   (2) The placements of emergency lay-bys and the cross passages for vehicles and pedestrians of new or expanded tunnels should be assessed.

5.4.5 Interchanges shall be audited in compliance with the requirements as follows:

1. The implications of the length from the gore nose to the start of the alignment control curve of a ramp, and of the radius of circular curve on traffic safety shall be evaluated.
2. The operating speed consistency for the ramp shall be evaluated. If the speed differential absolute values between operating speeds at successive geometric feature segments, or the speed differential between the predicted value of operating speeds at the ramp curve and the design speed of the ramp is greater than 20 km/h, the design consistency is poor.
3. Sight distance shall be evaluated in compliance with the requirements as follows:

   (1) The sight distance of the ramp typical sections should be evaluated according to the operating speed.
   (2) The visibility at nose of diverging areas and merging areas shall be evaluated according to the operating speed.

4. The evaluation of ramp entrances and exits shall be in compliance with the requirements as follows:

   (1) The distance between adjacent exits or entrances of main route or ramps, or from one exit to the next entrance on a main route shall be evaluated according to the operating speed.

(2) The length of acceleration and deceleration lanes shall be evaluated according to the operating speed of the main route, number of ramp lanes and gradient of the main route.

5. The operating speed-based design consistency along ramps of the highway upgrading and reconstruction projects should be evaluated.

5.4.6 Intersections shall be audited in compliance with the requirements as follows:

1. Speed-change lanes and auxiliary turning lanes at the intersections should be evaluated.

   (1) The length, width and gradient of speed-change lanes and geometric design criteria of transition sections such as the width and length should be evaluated according to the operating speeds and intersecting angles, etc.
   (2) The length of left-turning auxiliary lanes and the radius of right-turning lanes should be evaluated according to the method traffic control of intersections and operating speeds.

2. The suitability of lane quantities for all directions within channelization design should be evaluated according to the highway class and traffic volume.
3. The method of traffic control at intersections should be assessed.

5.4.7 Traffic control devices and roadside facilities shall be audited in compliance with the requirements as follows:

1. Traffic signs shall be evaluated in compliance with the requirements as follows:

   (1) The positioning of signs shall be assessed. The distance from warning signs to hazards should be evaluated according to the operating speed.
   (2) The rationality of information on signs, the continuity and effectiveness of information, as well as information load of directional signs shall be evaluated.
   (3) The dimensions of traffic signs, font and letter height shall be evaluated according to the operating speed.
   (4) The retroreflective class of traffic signs and compatibility with ambient lighting, climatic conditions, and operating speeds should be assessed.
   (5) The support of traffic signs shall be assessed according to the number of lanes, traffic composition and sign positioning.
   (6) For traffic signs within the suggested roadside clear zone, protection facilities for their foundation and mounting posts shall be considered.
   (7) The consistency of expressing the same information contents between traffic signs and road markings shall be assessed.

2. Road markings shall be evaluated in compliance with the requirements as follows:

   (1) The width, forms, colors, and retroreflective classes of road markings shall be assessed.

(2) Object markers of bridge piers and abutment, tunnel portals, sign posts shall be assessed.
(3) The location and length of deceleration markings or speed reduction pavement shall be evaluated.
(4) Rumble strips or raised pavement markers on the edge of traveled ways should be considered.
(5) The location and spacing of raised pavement markers shall be assessed.

3. The location and spacing of visual guiding devices such as delineators and alignment guiding device shall be evaluated.
4. Highway barriers shall be evaluated in compliance with the requirements as follows:

(1) If there are hazardous road sections such as high embankments, roadside close to water bodies or cliffs or villages, sections parallel to other roads, railways, oil and gas pipelines, steep grade and sharp curve sections, the protection level and type of adopted highway barriers shall be evaluated.
(2) The start and end, minimum length and spacing of barrier, and terminal treatment shall be evaluated.
(3) The connection and transition section between structures such as bridges and tunnels and connecting roads as well as the connection and transition section design between barriers of different rigidity shall be evaluated.
(4) The protection level and type of median barriers shall be evaluated according to the width of central dividing strips, traffic composition, operating speed and geometric conditions such as steep gradients and sharp curves.
(5) The protection level, type and position of median opening barriers shall be evaluated.
(6) If there are fixed rigid objects such as bridge piers (column) on the road or the median, the type of highway barriers shall be evaluated.
(7) For road sections and bridges provided with non-motorized lanes and sidewalks, facilities for segregation with the carriageway should be considered.

5. The implications of layout design and traffic management of service facilities, lanes and parking facilities within service areas and parking areas on traffic safety should be assessed.
6. For road sections with bus stops, their position and the length of acceleration or deceleration lanes should be evaluated.
7. For road sections with one or more features such as continuous upgrades, continuous long and steep downgrades, long downgrades preceding small-radius curves, long and large tunnels, tunnel-chains, bridge-tunnel sections, tunnel-interchange sections, sections susceptible to adverse weather and sections with severe roadside interference and roadside sheer drops, the integrated design of traffic control devices and roadside facilities shall be evaluated.
8. The evaluation of highway upgrading and reconstruction projects shall be in compliance with the requirements as follows:

(1) If existing traffic safety devices are reused in highway upgrading and reconstruction projects, their type and safety performance shall be assessed.
(2) For road sections with the separation of traffic in the same direction and transition sections between sections with different forms of widening, traffic safety devices and surveillance facilities shall be assessed.
(3) Lighting installed on transition sections at the start of separation of traffic in the same direction and on sections approaching interchanges with openings for traffic in the same direction should be considered.
(4) If new service facilities are provided for highway upgrading and reconstruction projects, the implications of the spacing between the new service facilities and other facilities or structures, as well as their entrances and exits on traffic safety shall be evaluated.

9. Speed limit proposals should be audited and be in compliance with the requirements as follows:

(1) The adopted types of speed limit should be assessed according to highway class, functions, traffic volume, traffic composition and width of cross-sections, etc.
(2) The posted speed limit should be assessed according to the alignment features and operating speed. Speed limit values for sections severely affected by alignment geometry, structures, roadside interference and adverse weather conditions should be a focus in the audit.

5.4.8 For highway upgrading and reconstruction projects, the design of traffic management shall be assessed according to the highway's technical class, traffic composition and traffic flow characteristics of the road network affected by the project in conjunction with the current operating conditions of the highway and the highway upgrading and reconstruction proposal.

## 5.5 Findings and Recommendations

5.5.1 The audit conclusions shall include findings and recommendations for overall audit and audit on design elements.

5.5.2 The conclusion of the overall audit and audit on design elements shall be in compliance with the requirements in Clauses 4.6.2, 4.6.4 and 4.6.5 of the *Specifications.*

5.5.3 Recommendations of safety improvements and countermeasures raised in the conclusions of the audit should place emphasis on the integrated provisions of traffic control devices and roadside facilities.

# Chapter 6
# Pre-opening Stage

## 6.1 General

6.1.1 This stage of audit shall focus on the placement of traffic control devices and roadside facilities during the pre-opening stage.

6.1.2 This stage audit shall consist of an overall audit and audit on highway safety performance, with the presumption that the quality of highway engineering has passed through acceptance tests and inspection.

6.1.3 The safety audit at the highway pre-opening stage shall be conducted according to the *Specifications*. The format of the *Highway Safety Audit Report* shall conform to the relevant requirements in Appendix A of the *Specifications*.

## 6.2 Methodology

6.2.1 A highway safety performance audit shall be carried out by on-site inspection and driving on the actual highway. Safety audit checklists and other methods should also be applied.

## 6.3 Overall Audit

6.3.1 The characteristics of the highway projects shall be analyzed and the implications on traffic safety shall be considered.



H. Wang, *Specifications for Highway Safety Audit*, JTG Standards and Specifications,
https://doi.org/10.1007/978-981-92-3132-4_6

6.3.2 The implementation of comments related to traffic safety shall be checked in the approval of detailed design.

6.3.3 If an audit has been carried out at the design stage, the actions in response to the comments of the safety audit shall be checked.

## 6.4 Audit on Highway Safety Performance

6.4.1 The route shall be audited in compliance with the requirements as follows:

1. The design continuity and design consistency of horizontal alignment and vertical alignment as well as the smoothness of cross-section transitions shall be reviewed using on-site driving tests.
2. The sight distance of horizontal alignment and vertical alignment shall reviewed using on-site driving tests.

6.4.2 Subgrade and pavement shall be audited in compliance with the requirements as follows:

1. Treatments of roadside obstructions shall be reviewed.
2. Drainage facilities of subgrades and pavements shall be reviewed.
3. The location and the sight distance of median openings shall be reviewed.

6.4.3 Bridges shall be audited in compliance with the requirements as follows:

1. For bridges preceded by sharp curve, relevant signs, road markings and speed management facilities shall be reviewed.
2. Transition sections at the connection between at-grade barriers and bridge railings shall be reviewed.
3. For bridges crossing over the project highway, the implications of piers, abutments and super-structures on sight distance of the highway shall be reviewed using on-site driving tests.
4. If the piers and abutments of bridges crossing over the audited project highway are located within the suggested clear zone, protection facilities for the bridge piers and abutments shall be considered.
5. Signs and speed management facilities shall be reviewed considering cross winds.

6.4.4 Tunnels shall be audited in compliance with the requirements as follows:

1. The consistency of alignment and sight distance at tunnel portals shall be reviewed using on-site driving tests.
2. Transitions of the skid resistance of pavements at tunnel portals shall be reviewed.
3. Transition sections at connections between the end of maintenance walkways (catwalks) and barriers outside the tunnel portal shall be reviewed.

4. The actual performance of tunnel lighting and glare conditions at tunnel portals shall be reviewed using on-site driving tests.
5. Provisions of tunnel surveillance and control facilities, ventilation facilities and fire-fight facilities should be reviewed.
6. For tunnels with mixed traffic, safety facilities for pedestrians and non-motorized vehicles shall be reviewed.

6.4.5 Interchanges shall be audited in compliance with the requirements as follows:

1. The visibility at the nose of a gore area or an entrance ramp (entire triangular area), the length of acceleration/deceleration lanes and operating speed consistency of ramps shall be reviewed using on-site driving tests.
2. Sign information at the entire exit terminal of interchanges shall be reviewed using on-site driving tests.

6.4.6 Intersections shall be audited in compliance with the requirements as follows:

1. The visibility of sight triangles shall be reviewed using on-site driving tests.
2. The method of traffic control and measures of traffic management shall be reviewed.
3. Signs, road markings and other traffic safety devices related to pedestrian and non-motorized vehicles shall be reviewed.

6.4.7 Traffic control devices and roadside facilities shall be audited in compliance with the requirements as follows:

1. The evaluation of traffic signs shall be in compliance with the requirements as follows:

   (1) The implications, performance and location of signs shall be reviewed on-site.
   (2) The accuracy and continuity of sign information shall be reviewed according to the road network conditions and on-site driving tests.
   (3) The amount of information on signs shall be reviewed.
   (4) The design consistency between information on signs and the corresponding road markings shall be reviewed.
   (5) The implications of trees, vegetation of slopes, structures and billboards on the visual recognition of traffic signs shall be evaluated.
   (6) Sign visibility and legibility at night shall be reviewed based on on-site driving tests.

2. Road markings shall be evaluated in compliance with the requirements as follows:

   (1) The visibility, legibility and guidance of road markings at night shall be reviewed using on-site driving tests.
   (2) Object markers for bridge piers, tunnel portals and supports of devices located in the median or the suggested clear zone shall be reviewed.

(3) The provision of road markings at no-overtaking shall be reviewed.

3. Highway barriers shall be evaluated in compliance with the requirements as follows:

   (1) The provision of highway barriers shall be reviewed.
   (2) Attenuation facilities at the noses of gore area shall be reviewed.
   (3) If there are hazardous road sections such as high embankments, roadside close to water bodies or cliffs or villages, sections parallel to other roads, railways, oil and gas pipelines, steep grade and sharp curve sections, the protection level and type of adopted highway barriers shall be reviewed.

4. Anti-glare devices shall be evaluated in compliance with the requirements as follows:

   (1) The provision of anti-glare devices shall be reviewed.
   (2) Anti-glare performance of anti-glare panels (glare screens) shall be reviewed at night.

5. The performance of the visual guiding devices shall be reviewed at night.
6. If the audited highway crosses railways, navigable rivers and other highways with heavy traffic volume, the adopted bridge nettings shall be reviewed.
7. Traffic safety devices and management facilities of climbing lanes and emergency escape ramps shall be reviewed.
8. The provision of surveillance and control facilities for the road section should be reviewed.
9. Traffic safety devices and administration facilities of toll stations shall be reviewed using on-site driving tests.
10. Signs and road markings in service areas and rest areas shall be reviewed.
11. Traffic safety devices at emergency lay-bys shall be reviewed.
12. For road sections with one or more features such as continuous upgrades, continuous long and steep downgrades, long downgrades preceding small-radius curves, long and large tunnels, tunnel-chains, bridge-tunnel sections, tunnel-interchange sections, sections susceptible to adverse weather and sections with severe roadside interference and roadside sheer drops, the integrated design of traffic control devices and roadside facilities shall be reviewed.
13. Speed limit proposals may be audited, focusing on posted speed limit values for road sections influenced by factors such as the alignment geometry, structures, roadside interference and meteorological conditions.

## 6.5 Findings and Recommendations

6.5.1 The audit conclusions shall include findings and safety recommendations for the overall audit and audit on highway safety performance.

6.5.2 The conclusions for the overall audit shall cover the characteristics of the highway project and their implications on traffic safety.

6.5.3 The audit conclusion on the highway safety performance shall cover the major issues which may affect traffic safety when the highway is in operation. Feasible recommendations for improvements shall be put forward based on the current status of the highway at the pre-opening stage.

6.5.4 Recommendations shall focus on the improvement of traffic safety devices or management countermeasures. The priority and phasing of recommendations should be clarified according to the level of difficulty of implementation.

# Chapter 7
# Operation Stage

## 7.1 General

7.1.1 The requirements in this Chapter are applicable to road safety audit in the highway post-construction stage. It is also applicable to road safety audit where there are distinct changes to the safety conditions of the highways after they are open to traffic or at the time of the as-built acceptance, major repair or medium rehabilitation, and highway upgrading and reconstruction projects.

7.1.2 The focus of the audit shall be on the implications of highway facilities, traffic volume and traffic composition, as well as network and roadside environment, etc. on traffic safety.

7.1.3 Overall audit and audit on highway safety performance shall be conducted.

7.1.4 Overall audit shall be based on research and data collection from the analyses of traffic crashes on the project highway. Audit on the highway safety performance shall include on-site inspection of the highway, speed observations and review of the safety recommendations and countermeasures.

7.1.5 The safety audit at the highway operation stage shall be conducted according to the Specification. The format of the *Highway Project Safety Audit Report* shall conform to the requirements in Appendix A of the *Specifications.*

## 7.2 Methodology

7.2.1 The overall audit should be based on methods such as statistical analyses of traffic crashes, questionnaire surveys, etc.



H. Wang, *Specifications for Highway Safety Audit*, JTG Standards and Specifications,
https://doi.org/10.1007/978-981-92-3132-4_7

7.2.2 The audit on highway safety conditions should use safety audit checklists and on-site observation of speeds, etc.

## 7.3 Overall Audit

7.3.1 The implications of traffic operation characteristics after a highway is open to traffic on traffic safety shall be evaluated according to traffic volume and composition, highway environment, safety management, climatic conditions and traffic crashes, etc.

7.3.2 Operational conditions, main causes of traffic crashes, road sections prone to traffic crashes and traffic safety management, etc. shall be investigated and studied.

7.3.3 Data shall be collected and their quality, quantity and timing shall meet the requirements for audit. Collection of information should include the following contents:

1. Current standards related to safety audit.
2. Statistical information such as traffic volume and traffic composition for the past three years or more.
3. Details of traffic crashes in the past three years or more, including the time, location, weather conditions, as well as the types, causes, number of casualties and vehicle types involved, etc.
4. Related design documents, including detailed design or as-built drawings of the project.
5. Information on traffic safety related to hand-over or as-built acceptance.
6. Other information which may be used for the safety audit.

7.3.4 Questionnaire surveys of highway users should be carried out, covering primarily the investigation of requirements for safe operation, effectiveness of safety administration measures and recommendations for safety improvements.

7.3.5 The analyses of traffic crashes shall be in compliance with the requirements as follows:

1. The statistics of the number of traffic crashes, number of casualties, economic losses, etc. shall be carried out. The trend of changes in traffic crashes shall be analyzed.
2. The statistical distribution of traffic crashes shall be analyzed and summarized, including the characteristics of traffic crashes distributed by time, space, type, causes and climatic conditions, etc.
3. Crash-prone sections shall be identified according to the spatial distribution of traffic crashes. The start and end of crash-prone sections shall be determined and the main causes of frequent crashes shall be analyzed.

4. Typical fatal and serious traffic crashes should be subject to a case-by-case analysis.

7.3.6 Highway facilities and emergency response plans related to EER may be reviewed.

## 7.4 Audit on Highway Safety Performance

7.4.1 On-site inspection for highway safety conditions shall be carried out. On-site inspection shall be in compliance with the requirements as follows:

1. Highway conditions shall be recorded by taking photographs or continuous video recording along the road in both driving directions.
2. Investigation of road sections prone to crashes shall be prioritized.
3. Road sections prone to crashes or to sections planned for speed management shall be subject to on-site observation of vehicles' speeds.
4. Class-1, Class-2 and Class-3 highways shall be investigated for roadside interference due to pedestrians and non-motor vehicles, etc.

7.4.2 Route shall be audited in compliance with the requirements as follows:

1. The speed differentials between design speeds and operating speeds shall be evaluated according to the operating speeds of representative vehicle types determined from data of on-site observations.
2. The continuity and sight distance of horizontal and vertical alignment shall be reviewed according to on-site driving tests.
3. For Class-2 and Class-3 highways, widening values of road sections with small-radii circular curves shall be reviewed according to the actual traffic composition.

7.4.3 Subgrades and pavements shall be audited in compliance with the requirements as follows:

1. The clearance profile of highway structures shall be reviewed.
2. Roadside obstacles within the suggested clear zone shall be reviewed.
3. For road sections prone to sideways skidding and rear-end collisions, the skid resistance of the pavement shall be evaluated.
4. The location and sight distance of median openings shall be reviewed.
5. The maintenance conditions and discharge capacity of drainage facilities shall be reviewed.

7.4.4 Bridges shall be audited in compliance with the requirements as follows:

1. The consistency of alignment between bridges and bridge approaches shall be reviewed. For bridges preceded by an approach on a sharp curve, relevant signs, road markings and speed management facilities shall be reviewed.

2. Transition sections at the connections between at-grade barriers and bridge railings shall be reviewed.
3. For bridges crossing over the project highway, the implications of piers, abutments and super-structures on sight distances of the highway shall be reviewed according to the on-site driving tests.
4. If the piers and abutments of bridges crossing over the audited highway project are located within the suggested clear zone, appropriate protection facilities for the bridge piers and abutments shall be reviewed.
5. The interface between bridge sections connecting with subgrade sections or locations with expansion joints shall be checked to determine whether traffic safety is affected by vehicle jumping effects.
6. Bridges shall be inspected for their susceptibility to slippery or icing conditions.
7. For bridges provided with non-motorized lanes and sidewalks, facilities for segregation from the carriageway shall be considered.
8. Signs and speed management facilities related to the cross-winds, etc. shall be reviewed.

7.4.5 Tunnels shall be audited in compliance with the requirements as follows:

1. The consistency of alignment and sight distance at tunnel portals shall be reviewed based upon the actual driving conditions present.
2. The skid resistance of pavements and transitions connecting the inside and outside of tunnel portals shall be reviewed.
3. Changes of cross-section at tunnel portals and connections and transition of protection facilities shall be reviewed.
4. The implications of ambient brightness and lighting transition at tunnel portals on traffic safety shall be reviewed using field driving conditions. The provisions of visual guiding devices shall also be reviewed for tunnels without lighting facilities.
5. The provisions of tunnel surveillance and control facilities, ventilation and fire-fight facilities, etc. should be reviewed.
6. Safety facilities for the protection of pedestrians and non-motorized vehicles shall be reviewed for tunnels with mixed traffic.

7.4.6 Interchanges shall be audited in compliance with the requirements as follows:

1. The spacing between adjacent interchanges, as well as between an interchange and adjacent service area, rest area, and tunnel, etc. shall be reviewed.
2. Visibility at the nose of a gore area or an entrance ramp (entire triangular area) shall be reviewed in accordance with operating speeds and using on-site driving tests.
3. The implications of the length from the nose of gore area to the start of the ramp curve and speed transition sections on traffic safety shall be reviewed in accordance with operating speeds and using on-site driving tests.
4. The balance in the number of lanes, length of speed changing lanes and auxiliary lanes and weaving sections shall be reviewed.

5. Sign information at the exit of interchanges shall be reviewed according to on-site driving tests and operating speeds.

7.4.7 The intersections shall be audited in compliance with the requirements as follows:

1. The implications of the location, form, crossing angle and spacing of intersections on traffic safety shall be reviewed through on-site observation.
2. The visibility condition of sight triangles shall be checked and reviewed in accordance with operating speeds.
3. Traffic safety facilities such as traffic channelization design, signs and road markings related to pedestrian and non-motor vehicles, etc. shall be reviewed.
4. Turning and additional lanes should be reviewed according to the class of intersecting highways, proportion of through traffic and turning traffic as well as historical traffic crashes, etc.
5. The adopted method of traffic control and traffic management measures should be reviewed in accordance with turning traffic volume and on-site conditions of the intersection.

7.4.8 The traffic control devices and roadside facilities shall be audited in compliance with the requirements as follows:

1. Signs, road markings, barriers, anti-glare devices, visual guiding devices and nettings shall be reviewed and be in compliance with the requirements in Clause 6.4.7 of the *Specifications*.
2. Administration facilities shall be evaluated in compliance with the requirements as follows:

   (1) The provisions of surveillance and lighting facilities on the road sections should be reviewed.
   (2) The traffic capacity and number of toll lanes, etc. at toll stations should be reviewed.
   (3) For toll stations preceded with a sharp curve or located at the bottom of continuous long downgrades, the sight distance, speed management facilities and protection facilities at toll booths, etc. shall be reviewed.

3. Service facilities shall be evaluated in compliance with the requirements as follows:

   (1) The decision sight distance of the nose of a gore area or an entrance ramp for the service areas and rest areas and the length of speed change lanes shall reviewed in accordance with the operating speeds.
   (2) For service areas and rest areas, their capacity, internal layout of lanes and internal service facilities, etc. should be reviewed according to actual traffic volume and traffic composition.
   (3) The location, speed change lanes and segregation facilities, etc. of bus stops shall be reviewed.

4. For road sections with one or more features such as continuous upgrades, continuous long and steep downgrades, long downgrades preceding small-radius curves, long and large tunnels, tunnel-chains, bridge-tunnel sections, tunnel-interchange sections, sections susceptible to adverse weather and sections with severe roadside interference and roadside sheer drops, the integrated design of traffic control devices and roadside facilities shall be reviewed.
5. If it is difficult for the speed-limit proposal to meet requirements for traffic safety or traffic flow efficiency, the proposal may be audited according to operational demand, highway conditions, operating speeds, traffic safety conditions, etc.

7.4.9 Control areas for maintenance works shall be audited in compliance with the requirements as follows:

1. The visibility of control areas for maintenance works and related safety devices shall be reviewed.
2. The implications of traffic management measures adopted over the duration of maintenance works on traffic safety should be considered.

## 7.5 Findings and Recommendations

7.5.1 The conclusions of the audit shall include the conclusions of the overall project and on the highway safety performance.

7.5.2 The conclusion of the overall project shall summarize the characteristics of the highways and their implications on traffic safety and analyze the causes of traffic crashes and trends of traffic safety.

7.5.3 The conclusion of the audit on the highway safety performance shall determine major safety issues and key areas of safety improvements, and propose feasible improvement proposals and management strategies.

7.5.4 The priority and phasing of implementation should be clarified according to the effectiveness of their implementation and the level of difficulty.

# Appendix A
# Format of Safety Audit Report

## A.1 Description of Report Format

A.1.1 Safety Audit reports should include the following contents:

1. Cover;
2. Qualification certificates;
3. Page of description;
4. Table of contents;
5. Texts.

A.1.2 *Safety Audit Reports* shall be of A4 size with binding on the left side.

A.1.3 The cover of *Safety Audit Reports* should be light gray.

A.1.4 The cover should consist of the following contents:

1. Name of the highway safety audit project;
2. Stages of audit;
3. Title, to be standardized as "Safety Audit Report"
4. Name of agency undertaking highway safety audit report;
5. Date of completing the audit report

The format of the cover is shown in Fig. A.1.4.

A.1.5 The description page should cover the following contents:

1. Name of the highway audit project;
2. Stage of audit;
3. Title, to be standardized as "Safety Audit Report"
4. Names of the undertaking agency including head, technical director, project manager and major participants;
5. Name of the undertaking agency and common seal or seal of technical achievements;



H. Wang, *Specifications for Highway Safety Audit*, JTG Standards and Specifications,
https://doi.org/10.1007/978-981-92-3132-4

**Name of the Audit Project**(Song typeface of font size 22 in bold)

**×× Stage**(Song typeface of font size 22 in bold)

**Safety Audit Report**

(Bold black Song typeface of font size 36)

**Name of agency undertaking highway safety audit report**

(Song typeface of font size 16 in bold)

**Date of completing the Audit report**

(Song typeface of font size 16 in bold)

**Fig. A.1.4** Cover format

6. Name and identification number of qualification certificates of the undertaking agency;
7. Date of completing the audit report

The format of the description page is shown in Fig. A.1.5.

## A.2 Format of the Main Body of Highway Safety Audit Report

A.2.1 The Text part of Safety Audit Report shall consist of following parts:

1. Overview

This is the description of the background, purpose, terms of reference and process as well the status of investigation and study of the safety audit.

2. Profile of the highway project

The part presents an overview of the construction works, physical and geographical conditions, traffic volume and composition, level of services, the relationship with other highways and railways along the route, specified and adopted parameters of major technical standards, as well as the main technical and economic data. At the pre-opening stage, the status of engineering construction and Hand-over acceptance, etc. shall also be illustrated. At the operation stage, the conditions of traffic management, safety improvements, and crashes shall be described.

**Name of Safety Audit Project**(Song typeface of font size 22 in bold)

**×× Stage**(Song typeface of font size 22 in bold)

**Safety Audit Report**

(Boldface Song typeface of font size 26)

**Head of Agency:**(Song typeface of font size 14 in bold)

**Technical director:**(Song typeface of font size 14 in bold)

**Project manager:**(Song typeface of font size 14 in bold)

**Main participants:**(Song typeface of font size 14 in bold)

**Name and seal of undertaking agency**

(Song typeface of font size 14 in bold)

**Name and identification number of qualification certificates of the undertaking agency**

(Song typeface of font size 14 in bold)

**Date of completing the audit report**

(Song typeface of font size 14 in bold)

**Fig. A.1.5** Format of the description page

3. Overall audit

The implications of the characteristics and environment of the highway project on traffic safety shall be analyzed in order to determine the focus of the audit. At the operation stage, it is required that the study results of highway operation and traffic crashes are analyzed.

4. Alternatives audit (preliminary design stage)

From the perspectives of traffic safety, alternative design proposals with the same level of details are compared and audit results are analyzed in order to recommend the preferred route proposal with the better performance of traffic safety.

5. The audit on design elements (stages of preliminary design and detailed design)

From the perspectives of traffic safety, design proposals of routes, subgrades, pavements, bridges, tunnels, interchanges, intersections, traffic engineering facilities and roadside facilities, etc. are audited.

6. Audit on highway safety performance (stages of Pre-opening and operation)

From the perspectives of traffic safety, the operating conditions of design proposals for routes, sub-grades, pavements, bridges, tunnels, interchanges, intersections, traffic engineering facilities and roadside facilities are inspected and reviewed.

7. Audit Conclusion and recommendations

# Appendix B
# Prediction Methods for Operating Speeds

## B.1 General

B.1.1 The representative vehicle types used for the forecast of highway operating speeds at the preliminary design stage and the detailed design stage shall comply with the requirements in Table B.1.1.

B.1.2 According to the value of the curve radii and longitudinal grade gradients, a highway may be divided into several analysis units; including tangent sections, sections of horizontal curve, sections on longitudinal grade, grade-curve sections (sections with a combination of horizontal curve and longitudinal gradient), tunnel sections and interchange sections, etc. Road sections of horizontal curve or combinations of horizontal curve and longitudinal gradient should be analyzed as separate sections around the mid-point of the horizontal curve.

B.1.3 The operating speeds $v_{85}$ of small vehicles and large vehicles shall be predicted separately for each driving direction. The start and end points of the analysis unit should be the feature points used for determining operating speeds $v_{85}$. The direction of longitudinal grade shall be consistent with the driving direction.

B.1.4 When making predictions of operating speed $v_{85}$, the initial operating speed, $v_0$, at the start point of the first analysis unit of the road section shall be determined first. The operating speed, at the end point of the first analysis unit is *then* calculated according to prediction models of operating speeds corresponding to the driving direction and the analysis unit, and this serves as the initial operating speed of the second analysis unit. Thereafter, the operating speed at the end point of the second

**Table B.1.1** Representative vehicles for operating speed prediction

| Vehicle types | Motorways and Class-1 highway | Class-2 and Class-3 highways |
|---|---|---|
| Small vehicles | Wheelbase ≤7 m and specific power > 15 kW/t | Wheelbase ≤3.5 m |
| Large vehicles | Wheelbase >7 m or specific power ≤ 15 kW/t | Wheelbase >3.5 m |



H. Wang, *Specifications for Highway Safety Audit*, JTG Standards and Specifications,
https://doi.org/10.1007/978-981-92-3132-4

analysis unit is calculated by substituting this value into the calculation formula for the second unit. The calculation is iterated using this method until the last analysis unit.

B.1.5 *Calculation* of the operating speed of analysis units shall be in compliance with the requirements as follows:

1. The initial operating speed, $v_0$, and the desired speed of the first analysis unit, $v_e$, should be determined by using the specified values in this Appendix or determined from observational results of similar highways in the project area. The desired speed in the *Specifications*, $v_e$, refers to the maximal driving speed expected by drivers on tangent sections (i.e., straight and flat sections) of the highway under ideal traffic conditions such as clear and fine weather, dry pavement, good conditions of highway and ancillary facilities, free-flow conditions and absence of interference.
2. Prediction models of operating speeds should be established with the setting of parameters according to observational results of similar highways in the project area. If there are constraints, the model of this Appendix may be used.
3. Prediction models of operating speeds consist of a basic model and calibration methods. The basic model includes models for tangent sections, horizontal curve sections, grade-curve sections, tunnel sections and interchange sections, as well as the conversion model of longitudinal grade sections. For analysis units in need of correction, corrections should be made on the basis of results from the basic model and the corrected results should be adopted as the final results of the calculation.
4. Operating speed calculated using prediction models of operating speeds shall be in compliance with the requirements as follows:

   (1) The operating speeds at the start point and end point of an analysis unit shall be neither greater than the desired speed, nor less than the minimum operating speed corresponding to the analysis unit or the minimum speed limit values.

   (2) If the calculation result is greater than or equal to the desired speed, should be used to represent the operating speed at the end point of the analysis unit. If the calculation result is less than the minimum operating speed corresponding to the analysis unit or the minimum speed limit value, the minimum operating speed for the analysis unit or the minimum speed limit value should be used to represent the operating speed at the end point of the analysis unit.

## B.2 Operating Speeds of Motorways

B.2.1 The division of analysis units should comply with the requirements as follows:

**Table B.2.1** Principles for classification of analysis unit

| Vehicle types | Vertical | Horizontal alignment | |
|---|---|---|---|
| | | Radius of circular curve>1000 m | Radius of circular curve ≤1000 m |
| Small or large vehicles | Grade < 3% | Flat tangent section if length > 200 m.<br>Short, flat tangent section if length ≤ 200 m | Horizontal curve section |
| | Grade ≥ 3% | Longitudinal grade section | Grade-curve section |

**Table B.2.2-1** Initial operating speed (km/h)

| Design speed | | 120 | 100 | 80 | 60 |
|---|---|---|---|---|---|
| Initial operating speed | Small vehicles | 120 | 100 | 80 | 60 |
| | Large vehicles | 80 | 75 | 65 | 50 |

1. The motorways should be divided into analysis units such as the tangent sections, horizontal curve sections, longitudinal grade sections, grade-curve sections, tunnel sections and interchange sections.
2. The division of tangent sections, longitudinal grade sections, horizontal curve sections, and grade-curve sections should comply with requirements in Table B.2.1.
3. Tunnel sections should cover a section of 200 m before drivers enter the tunnel and 100 m after drivers exit the tunnel.
4. Interchange sections should cover the section from the start of taper of the deceleration lanes to the end of taper of the acceleration lane. Ramp sections should cover the section from the connecting point with the main route to the end point of the ramp.

B.2.2 The initial operating speed, the desired speed, the minimum operating speed and the acceleration of small or large vehicles should comply with the requirements as follows:

1. The initial operating speed should be determined according to design speed in Table B.2.2-1.
2. The desired speed should be determined as per Table B.2.2-2.
3. The recommended acceleration should be determined as per Table B.2.2-3.
4. The minimum operating speeds of the small and large vehicles should not be less than 50 km/h and 30 km/h respectively.

B.2.3 The operating speed of tangent sections should comply with the requirements as follows:

1. If the length of the divided tangent section is more than 200 m, the operating speed model at the end point of the tangent section should be determined according to Eq. (B.2.3-1).

**Table B.2.2-2** Desired speed (km/h)

| Design speed | | 100 or 120 | 80 | 60 |
|---|---|---|---|---|
| Desired speed | Small vehicles | 120 | 110 | 90 |
| | Large vehicles | 80 | 80 | 75 |

**Table B.2.2-3** Recommended acceleration ($m/s^2$)

| Vehicle type | $a_{min}$ | $a_{max}$ |
|---|---|---|
| Small vehicles | 0.15 | 0.50 |
| Large vehicles | 0.20 | 0.25 |

$$v_{out} = 3.6\sqrt{\left(\frac{v_{in}}{3.6}\right)^2 + 2as} \quad \text{(B.2.3-1)}$$

Where:

$v_{out}$ Speed at the end point of the tangent section (km/h);
$v_{in}$ Speed at the start point of the tangent section (km/h);
$s$ Length of the tangent section (m);
$a$ Acceleration of vehicles ($m/s^2$) calculated according to Eq. (B.2.3-2):

$$a = a_{min} + (a_{max} - a_{min})\left(1 - \frac{v_{in}}{v_e}\right) \quad \text{(B.2.3-2)}$$

$a_{max}$ Maximal acceleration ($m/s^2$);
$a_{min}$ Minimal acceleration ($m/s^2$);
$v_e$ Desired speed (km/h).

2. If the length of a divided tangent section is not more than 200 m, such sections should be regarded as the short flat tangent section. The operating speeds at the start and end points of the road section remain constant.

B.2.4 The prediction of operating speeds at horizontal curve sections should comply with the requirements as follows:

1. The form of connections of horizontal curves, consisting of the entrance straight, tangent-curve, entrance curve-curve, exit curve-straight tangent and exit curve-curve should be determined.
2. The operating speeds at the midpoint and exit of a curve should be predicted based on segmentation at the midpoint of the curve.
3. The operating speeds at the midpoint and exit of a curve should be predicted according to the model in Table B.2.4.

**Table B.2.4** Prediction models of operating speeds on horizontal curve sections

| Connecting form of horizontal curves | Vehicle type | Prediction model |
|---|---|---|
| Entrance tangent-curve | Small vehicle | $v_{\text{middle}} = -24.212 + 0.834\ v_{\text{in}} + 5.729\ln R_{\text{now}}$ |
| | Large vehicle | $v_{\text{middle}} = -9.432 + 0.963\ v_{\text{in}} + 1.522\ln R_{\text{now}}$ |
| Entrance tangent-curve | Small vehicle | $v_{\text{middle}} = 1.277 + 0.942\ v_{\text{in}} + 6.19\ln R_{\text{now}}\text{-}5.959\ln R_{\text{back}}$ |
| | Large vehicle | $v_{\text{middle}} = -24.472 + 0.990\ v_{\text{in}} + 3.629\ln R_{\text{now}}$ |
| Exit curve-straight tangent | Small vehicle | $v_{\text{out}} = 11.946 + 0.908\ v_{\text{middle}}$ |
| | Large vehicle | $v_{\text{out}} = 5.217 + 0.926\ v_{\text{middle}}$ |
| Exit curve-curve | Small vehicle | $v_{\text{out}} = -11.299 + 0.936\ v_{\text{middle}}\text{-}2.060\ln R_{\text{now}} + 5.203\ln R_{\text{front}}$ |
| | Large vehicle | $v_{\text{out}} = 5.899 + 0.925\ v_{\text{middle}}\text{-}1.005\ln R_{\text{now}} + 0.329\ln R_{\text{front}}$ |

*Note*

1. $v_{\text{middle}}$: In the table refers to the operating speed at the midpoint of the curve;
2. $v_{\text{out}}$: Operating speed of the exit of the curve (km/h);
3. $R_{\text{front}}$: Radius of the curve to be entered (km/h);
4. $R_{\text{now}}$: Radius of the current curve (m);
5. $R_{\text{back}}$: Radius of the curve before entering the current curve.

**Table B.2.5** Conversion model of operating speed at longitudinal grade sections

| Longitudinal grade | | Adjustment value of operating speed | |
|---|---|---|---|
| | | Small vehicles | Large vehicles |
| Uphill | Gradient≥3% and ≤ 4% | Reduction of 5 km/h per 1, 000 m until the minimum operating speed is reached | Reduction of 10 km/h per 1, 000 m until the minimum operating speed is reached |
| | Gradient>4% | Reduction of 8 km/h per 1, 000 m until the minimum operating speed is reached | Reduction of 20 km/h per 1, 000 m until the minimum operating speed is reached |
| Downhill | Gradient≥3% and ≤ 4% | Increase of 10 km/h per 500 m until the desired speed is reached | Increase of 7.5 km/h per 500 m until the desired speed is reached |
| | Gradient>4% | Increase of 20 km/h per 500 m until the desired speed is reached | Increase of 15 km/h per 500 m until the desired speed is reached |

B.2.5 The operating speed of the end point of longitudinal grade section should be converted in accordance with the model of Table B.2.5.

B.2.6 The prediction of operating speeds at grade-curves sections should comply with the requirements as follows:

**Table B.2.6** Prediction models of operating speeds at slope-curve sections

| Form of slope-curve | Vehicle type | Prediction model |
|---|---|---|
| Entrance tangent-curve | Small vehicle | $v_{middle} = -31.67 + 0.547\ v_{in} + 11.71\ln R_{now}$-$0.176 I_{now1}$ |
| | Large vehicle | $v_{middle} = 1.782 + 0.859\ v_{in}$-$0.51 I_{now1} + 1.196\ln R_{now}$ |
| Entrance curve-curve | Small vehicle | $v_{middle} = 0.750 + 0.802\ v_{in} + 2.717\ln R_{now}$-$0.281 I_{now1}$ |
| | Large vehicle | $v_{middle} = 1.798 + 0.248\ln R_{now} + 0.977\ v_{in}$- $0.133 I_{now1} + 0.23\ln R_{back}$ |
| Exit curve-tangent | Small vehicle | $v_{out} = 27.294 + 0.720\ v_{middle}$-$1.444 I_{now2}$ |
| | Large vehicle | $v_{out} = 13.490 + 0.797\ v_{middle}$-$0.6971 I_{now2}$ |
| Exit curve-curve | Small vehicle | $v_{out} = 1.819 + 0.839\ v_{middle}$+$1.427\ln R_{now} + 0.782\ln R_{front}$-$0.48 I_{now2}$ |
| | Large vehicle | $v_{out} = 26.837 + 0.109\ln R_{front}$-$3.039\ln R_{now}$-$0.594 I_{now2} + 0.830\ v_{middle}$ |

*Note*

2. $v_{in}$, $v_{middle}$ and $v_{out}$ represent the operating speeds at the entrance, midpoint and exit of the curve.
3. $R_{back}$, $R_{now}$ and $R_{front}$ represent the curve radius before entering the curve, the radius of the current curve, the radius of the curve to be entered.
4. $I_{now1}$ and $I_{now2}$ represent two different slope grades at the front and rear sections of the curve. Uphill grade is positive and downhill grade is negative. The grade values with the plus or minus signs, without the percentage are substituted into a formula. For example, the uphill "4%" is substituted by value "4" and the downhill "−4%" is substituted by value "−4".
5. If the first half or the latter half section contains two different values of longitudinal grade, the weighted average of the longitudinal grade shall be substituted into the formula.

1. The form of grade-curve sections should be determined based on the type of connection of successive alignments. These types include entrance tangent-curve, entrance curve-curve, exit curve-tangent and exit curve-curve.
2. The operating speeds at the midpoint and exit of curves should be predicted with segmentation at the midpoint of the circular curve.
3. The operating speeds at the midpoint and exit of curves should be predicted according to the model in Table B.2.6.

B.2.7 The operating speed at tunnel sections should be predicted according to the model in Table B.2.7.

B.2.8 The prediction of operating speeds at the interchange sections should comply with the requirements as follows:

**Table B.2.7** Prediction models of operating speeds on tunnel sections

| Vehicle type | Feature point | Prediction model |
|---|---|---|
| Small vehicles | Tunnel entrance portal | $v_1 = 0.99\ v_{in}$-11.07 |
| | In the tunnel | $v_2 = 0.81\ v_{in} + 8.22$ |
| | 100 m after driving out of the tunnel exit portal | $v_3 = 0.74\ v_{in} + 16.43$ |
| Large vehicles | Entrance portal | $v_1 = 0.98\ v_{in}$-6.56 |
| | In the tunnel | $v_2 = 0.85\ v_{in} + 3.89$ |
| | 100 m after driving out of the tunnel exit portal | $v_3 = 0.45\ v_{in} + 42.61$ |

*Note* $v_{in}$ in the table represents the driving speed at the connecting road section 200m away from the tunnel portal to be entered (km/h). Except for short tunnels which are calculated as the short flat tangent sections, other tunnels are calculated according to the above model. For chains of tunnels with spacing of less than 200 m between an exit and an entrance of successive tunnels, may be regarded as the same tunnel section.

**Table B.2.8** Reduction value of operating speeds at mainline sections of interchanges

| Vehicle type | Small vehicle | Large vehicle |
|---|---|---|
| Maximal reduction value | −8 | −5 |

1. The operating speeds on road sections of the main route should be reduced according to Table B.2.8, on the basis of prediction of operating speeds before classification of analysis units for the interchange sections.
2. The operating speed at ramp sections should be determined according to observation of similar highways in the project area or implemented in accordance with relevant requirements in Clauses B.4.5-B.4.8 of the *Specifications.* The speed passing the gore nose of diverging areas should be adopted as the initial operating speed of ramp sections.

## B.3 Operating Speed on Class-1 Highways

B.3.1 Predictions of operating speeds on Class-1 highways serving as arterial highways should comply with the requirements as follows:

1. For road sections without intersections, without roadside interference or resembling full access controlled operation, prediction of operating speeds may be implemented in accordance with the requirements of motorways stipulated in Section B.2 of the *Specifications.*
2. For road sections with intersections, roadside interference or resembling sections with partial access control, efforts should be made to observe the implications of

intersections on similar highways in the project area in order to amend the prediction results of operating speeds. Otherwise, these are determined in accordance with the requirements in Clause B.3.2 of the *Specifications*.

B.3.2 The prediction of operating speeds of collector-distributor Class-1 highways should comply with the requirements as follows:

1. The operating speeds of analysis units are predicted in accordance with the relevant requirements in Section B.2 of the *Specifications* for motorways.
2. The operating speeds of analysis units should be corrected according to the requirements in Clauses B.3.3 and B.3.4 of the *Specifications*, based on the conditions of roadside interference and density of intersections.

B.3.3 The operating speed of analysis units should be corrected according to conditions of the roadside interference and shall comply with the requirements as follows:

1. In terms of roadside interference, the interference caused to small and large vehicles on the mainline by the flow of pedestrians, non-motor vehicles, motorcycles and agricultural vehicles, etc. shall be analyzed.
2. The level of roadside conflicts should be used to quantify the severity of roadside conflicts. The implications of roadside conflicts on operating speeds are determined from Table B.3.3-1. Operating speeds should be corrected by multiplying the operating speed value by the corresponding influence coefficient as shown in Table B.3.3-1 based on the roadside conflict levels of the analysis unit.
3. Roadside conflict level should be determined from the corresponding weighted values of roadside conflict factor, FRIC, in Table B.3.3-2.
4. The weighted value of roadside conflicts, FRIC, represents the weighted value of roadside conflict occurred per unit of time within 200 m of the observational section. FRIC may be determined according to Eq. (B.3.3-1).

$$\mathrm{FRIC} = 0.129 \cdot bic + 0.164 \cdot psv + 0.185 \cdot tra + 0.148 \cdot ped + 0.171 \cdot smv + 0.202 \cdot mot \quad \text{(B.3.3-1)}$$

Where:

*bic* Number of bicycle (vehicle/h);
*psv* Number of parked vehicles at the roadside (vehicle/h);
*tra* Number of slow-moving vehicles including tractors and other agricultural vehicles (vehicle/h);
*ped* Number of pedestrian (person/h);
*smv* Number of electric bicycle (vehicle/h);
*mot* Motorcycle (vehicle/h).

B.3.4 The operating speed of analysis units should be corrected according to the equivalent intersection density and should comply with the requirements as follows:

**Table B.3.3-1** Table of influence coefficient of roadside interference level on operating speeds

| Roadside interference level | Influence coefficient | Roadside conflict level | Influence coefficient |
|---|---|---|---|
| 0 | 1.00 | 2 | 0.82 |
| 1 | 0.91 | 3 | 0.73 |

**Table B.3.3-2** Table of roadside conflict level

| Roadside interference level | FRIC | Nature of land use on both sides of the highway |
|---|---|---|
| 0 | $0 < \text{FRIC} \leq 50$ | There are farmlands or mountains, canyons, etc. |
| 1 | $50 < \text{FRIC} \leq 100$ | There are sparse farmhouses and a small number of pedestrians |
| 2 | $100 < \text{FRIC} \leq 150$ | There is a small number of pedestrians and turning vehicles, presence of gas stations and small shops, etc. |
| 3 | $\text{FRIC} \geq 150$ | There is a high level of urbanization, featured by residential areas, commercial centers and a lot of pedestrians and vehicles |

**Table B.3.4-1** Table for implications of influence coefficient of equivalent intersection density on operating speeds

| Equivalent intersection density (no./km) | Influence coefficient | | |
|---|---|---|---|
| | $V_{85} \geq 100$ km/h | $80 \leq V_{85} < 100$ km/h | $60 \leq V_{85} < 80$ km/h |
| $0 < d \leq 1.0$ | 0.99 | 0.99 | 1.00 |
| $1.0 < d \leq 2.5$ | 0.98 | 0.98 | 0.99 |
| $2.5 < d \leq 5.0$ | 0.95 | 0.96 | 0.97 |
| $d > 5.0$ | 0.90 | 0.92 | 0.94 |

1. The implications of equivalent intersection density on operating speed should be determined according to Table B.3.4-1. Operating speeds should be corrected by multiplying the density value by the corresponding influence coefficient in Table B.3.4-1 based on the equivalent intersection density of the analysis unit.
2. The equivalent intersection density is calculated by dividing the sum of the number of equivalent intersections in the analysis unit by the length of analysis unit.
3. The number of equivalent intersections should be converted depending upon the traffic volume of motor vehicles at the intersections according to Table B.3.4-2.

**Table B.3.4-2** Conversion coefficient of intersections

| Traffic volume of motor vehicles on the intersecting road (veh/h) | Conversion coefficient of intersections |
|---|---|
| Vol $\leq$ 30 | 0.5 |
| 30 < vol $\leq$ 70 | 1.0 |
| 70 < vol $\leq$ 150 | 2.0 |
| Vol >150 | 3.0 |

**Table B.4.2** Principles for classification of analysis units

| Vehicle type | Vertical section | Plane | |
|---|---|---|---|
| | | Radius of circular curve greater than 600 m | Radius of circular curve not greater than 600 m |
| Small or large vehicle | Grade < 3% | Tangent section with length greater than 100 m; Short tangent section with length less than 100 m | Horizontal curve section |
| | Grade $\geq$ 3% | Longitudinal grade section | Grade-curve section |

## B.4 Operating Speeds on Class-2 and Class-3 Highways

B.4.1 The prediction of operating speeds on Class-2 and Class-3 highways and interchange ramps with design speed of 40 km/h and above should comply with the relevant requirements in this Section. For highways with design speed less than 40 km/h, models should be established or prediction models of operating speeds in this Section should be corrected according to the conditions of similar highways in the project area.

B.4.2 The division of analysis units should comply with the requirements as follows:

1. Highways should be divided into tangent sections, longitudinal grade sections, horizontal curve sections, grade-curve sections, tunnel sections, etc.
2. The classification of tangent sections, longitudinal grade section, horizontal curve section, and grade-curve section should comply with requirements in Table B.4.2.
3. A tunnel section should include a distance of 200 m before driving into the tunnel portal and 100 m after driving out of the tunnel portal.

B.4.3 The prediction of the operating speed of analysis units should comply with the requirements as follows:

1. Operating speeds of tangent sections, longitudinal grade sections, horizontal curve sections, and grade-curve sections should be predicted according to the corresponding models in B.4.5-B.4.8 of the *Specifications.*
2. Operating speeds of analysis units such as tangent sections, longitudinal grade sections, horizontal curve sections, and grade-curve sections should be corrected

**Table B.4.4-1** Initial operating speed (km/h)

| Design speed | | 80 | 60 | 40 |
|---|---|---|---|---|
| Initial operating speed, $v_0$ | Small vehicle | 80 | 60 | 40 |
| | Large vehicle | 60 | 40 | 30 |

**Table B.4.4-2** Desired speed (km/h)

| Design speed (km/h) | | 80 | 60 | 40 |
|---|---|---|---|---|
| Desired speed, $v_e$ | Small vehicle | 105 | 85 | 65 |
| | Large vehicle | 75 | 70 | 50 |

**Table B.4.4-3** Recommended acceleration ($m/s^2$)

| Vehicle types | $a_{min}$ | $a_{max}$ |
|---|---|---|
| Small vehicle | 0.15 | 0.50 |
| Large vehicle | 0.20 | 0.25 |

with respect to the width of the roadside clear zone, the density of access and conditions of roadside interference, according to Clauses B.4.9-B.4.11 of the *Specifications.*

3. The operating speeds of tunnel sections should be determined by actual observational results for similar highways in the project area or calculated according to the design speed of the highway.

B.4.4 The initial operating speed, the desired speed, the minimum operating speed and the acceleration of the small or large vehicles should comply with the requirements as follows:

1. The initial operating speed, $v_0$, should be determined in accordance with Table B.4.4-1.
2. $v_e$, the desired speed, should be determined in accordance with Table B.4.4-2.
3. The recommended acceleration, $a$, should be determined in accordance with Table B.4.4-3.
4. The minimal operating speed of small vehicles should not be less than 30 km/h and the minimal operating speed of large vehicles should not be less than 15 km/h.

B.4.5 Prediction of operating speeds on tangent sections should comply with the requirements in Clause B.2.3 of the *Specifications.*

B.4.6 The prediction of operating speeds on horizontal curve sections should comply with the requirements as follows:

1. The operating speeds at the midpoint and exit of curves should be predicted based on the segmentation from the midpoint of the curve.

**Table B.4.6** Prediction models of operating speeds on horizontal curve sections

| Feature points | Vehicle type | Prediction model |
|---|---|---|
| Midpoint of curve | Small vehicles | $v_{\text{middle}} = -244.123 + 0.6\ v_{\text{in}} + 40\ln\,(R_{\text{now}} + 500)$ |
| | Large vehicles | $v_{\text{middle}} = -80.179 + 0.7\ v_{\text{in}} + 15\ln\,(R_{\text{now}} + 250)$ |
| Exit of curve | Small vehicles | $v_{\text{out}} = -183.092 + 0.7\ v_{\text{middle}} + 30\ln\,(R_{\text{front}} + 500)$ |
| | Large vehicles | $v_{\text{out}} = -53.453 + 0.8\ v_{\text{middle}} + 10\ln\,(R_{\text{front}} + 250)$ |

*Note*

1. In the table $v_{\text{in}}$ represents the operating speed at the curve entrance; $v_{\text{middle}}$ represents the operating speed at the midpoint of the curve and $v_{\text{out}}$ represents the operating speed at the exit of the curve;
2. $R_{\text{now}}$: the radius of the current curve; $R_{\text{front}}$: the radius of the curve to be entered, if there is a tangent ahead, $R_{\text{front}} = 600$ m;
3. For small vehicles, if $R_{\text{front}} > 5R_{\text{now}}$ the values are taken with $R_{\text{front}} = 5R_{\text{now}}$; for large vehicles, if $R_{\text{front}} > 4R_{\text{now}}$ the values are taken with $R_{\text{front}} = 4R_{\text{now}}$.

**Table B.4.7** Model for conversion of operating speeds on longitudinal grade sections

| Longitudinal grade | | Adjustment values of operating speeds | |
|---|---|---|---|
| | | Small vehicles | Large vehicles |
| Uphill | Gradient ≥3% and ≤ 4% | Reduction of 5 km/h per 1, 000 m until the minimum operating speed is reached | Reduction of 10 km/h per 1, 000 m until the minimum operating speed is reached |
| | Gradient >4% | Reduction of 8 km/h per 1, 000 m until minimum operating speed is reached | Reduction of 20 km/h per 1, 000 m until the minimum operating speed is reached |
| Downhill | Gradient ≥3% and ≤ 4% | Increase of 10 km/h per 500 m until the desired speed is reached | Increase of 7.5 km/h per 500 m until the desired speed is reached |
| | Gradient >4% | Increase of 20 km/h per 500 m until the desired speed is reached | Increase of 15 km/h per 500 m until the desired speed is reached |

2. Operating speeds at the midpoint and exit of curves should be predicted according to the model in Table B.4.6.

B.4.7 Operating speeds at the end point of longitudinal grade sections should be predicted in accordance with the model of Table B.4.7.

B.4.8 Operating speeds on grade-curve sections should comply with the requirements as follows:

1. The prediction of operating speeds at the midpoint and exit of curves respectively on grade-curve sections should be based on the segmentation from the midpoint of a circular curve.
2. The form of grade-curve connections should be determined by the transition mode of the preceding and succeeding alignments. These forms include uphill

or downhill on the first half of the curved grade, as well as uphill or downhill on the second half on the curved grade.

3. The operating speeds at the midpoint and exit of curves on grade-curve sections should be predicted according to the model in Table B.4.8.

B.4.9 The implications of the roadside clear zone distance on operating speeds should be determined in accordance with Table B.4.9. The results of operating speeds should

**Table B.4.8** Prediction models of speeds on grade-curve sections

| Form of slope-curve | | Vehicle type | Prediction model |
|---|---|---|---|
| Midpoint of curve | Uphill on the first half of the section | Small vehicle | $v_{middle} = -244.123 + 0.6\, v_{in} + 40\ln(R_{now} + 500)\text{-}\frac{(600-R_{now})(i_1-3)}{600}\text{-}0.324 i_2$ |
| | | Large vehicle | $v_{middle} = -80.179 + 0.7\, v_{in} + 15\ln(R_{now} + 250)\text{-}\frac{1.2(600-R_{now})(i_1-2)}{600}\text{-}0.106 i_2$ |
| | Downhill on the first half of the section | Small vehicle | $v_{middle} = -244.123 + 0.6\, v_{in} + 40\ln(R_{now} + 500)\text{-}\frac{0.6R_{now}(i_1+3)}{600}\text{-}0.324 i_2$ |
| | | Large vehicle | $v_{middle} = -80.179 + 0.7\, v_{in} + 15\ln(R_{now} + 250)\text{-}\frac{0.8R_{now}(i_1+2)}{600}\text{-}0.106 i_2$ |
| Exit of curve | Uphill on thc latter half of the section | Small vehicle | $v_{out} = -183.092 + 0.7\, v_{middle} + 30\ln(R_{front} + 500)\text{-}\frac{1.2(600-R_{now})(i_2-3)}{600}\text{-}0.324 i_3$ |
| | | Large vehicle | $v_{out} = -53.453 + 0.8\, v_{middle} + 10\ln(R_{front} + 250)\text{-}\frac{1.5(600-R_{now})(i_2-2)}{600}\text{-}0.106 i_3$ |
| | Downhill on the second half of the section | Small vehicle | $v_{out} = -183.092 + 0.7\, v_{middle} + 30\ln(R_{front} + 500)\text{-}\frac{0.8R_{now}(i_2+3)}{600}\text{-}0.324 i_3$ |
| | | Large vehicle | $v_{out} = -53.453 + 0.8\, v_{middle} + 10\ln(R_{front} + 250)\text{-}\frac{R_{now}(i_2+2)}{600}\text{-}0.106 i_3$ |

*Note*

1. In the table $v_{in}$ represents the operating speed at the curve entrance; $v_{middle}$ represents the operating speed at the curve midpoint; $v_{out}$ represents the operating speed at the exit of the curve.
2. $R_{now}$: the radius of the current curve; $R_{front}$: the radius of the curve to be entered, if there is a tangent ahead, $R_{front} = 600$ m.
3. For small vehicles, if $R_{front} > 5R_{now}$, the values are taken with $R_{front} = 5R_{now}$; for large vehicles, if $R_{front} > 4R_{now}$, $R_{front} = 4R_{now}$.
4. $i_1$: the longitudinal grade before the midpoint of the slope-curve (%); $i_2$: the longitudinal grade after the midpoint of the slope-curve (%); $i_3$: the longitudinal grade before the slope-curve (%).
5. If the calculation results of $v_{middle}$ and $v_{out}$ are less than the minimum operating speed, the values are taken according to the minimum operating speed; if they are greater than the desired speed, values are taken according to the desired speed value.
6. If the first half or the second half contains two different values of longitudinal grades, the weighted average value of gradients of longitudinal grades should be substituted into the formula.

**Table B.4.9** Table for influence coefficient of roadside clear zone on operating speeds

| Width of clear zone (m) | 0.5 | 0.75 | 1.00 | 1.25 | 1.5 | 1.75 | 2 | 2.5 |
|---|---|---|---|---|---|---|---|---|
| Influence coefficient | 0.88 | 0.93 | 0.97 | 1.00 | 1.02 | 1.04 | 1.06 | 1.09 |
| Width of clear zone (m) | 3 | 4 | 5 | 6 | 7 | 8 | 9 | |
| Influence coefficient | 1.11 | 1.15 | 1.17 | 1.20 | 1.22 | 1.23 | 1.25 | |

**Table B.4.10** Table for influence coefficients of intersection density on operating speeds

| Intersection Density (number/km) | Influence coefficient | | | | |
|---|---|---|---|---|---|
| | 90 km/h | 80 km/h | 70 km/h | 60 km/h | 50 km/h |
| 5.0 | 0.89 | 0.92 | 0.94 | 0.96 | 0.97 |
| 2.5 | 0.93 | 0.94 | 0.96 | 0.97 | 0.98 |
| 2.0 | 0.94 | 0.94 | 0.96 | 0.98 | 0.98 |
| 1.0 | 0.97 | 0.97 | 0.98 | 0.99 | 0.99 |
| 0.5 | 0.98 | 0.99 | 0.99 | 0.99 | 0.99 |
| 0.3 | 0.99 | 1.00 | 1.00 | 1.00 | 1.00 |

be corrected by multiplying the width of roadside clear zone of the analysis unit by the influence coefficient in Table B.4.9.

B.4.10 The implications of intersection density on operating speeds should be determined in accordance with Table B.4.10. The density of intersections is calculated by dividing the sum of the number of intersections by the length of the analysis unit. With respect to the density of intersections, the results of operating speeds should be corrected by multiplying the density value by the corresponding influence coefficient in Table B.4.10.

B.4.11 The results of operating speeds should be corrected with the influence coefficients in Figs. B.4.11-1, B.4.11-2 and B.4.11-3 based on the number of roadside interference features and the lateral distance from roadside interference.

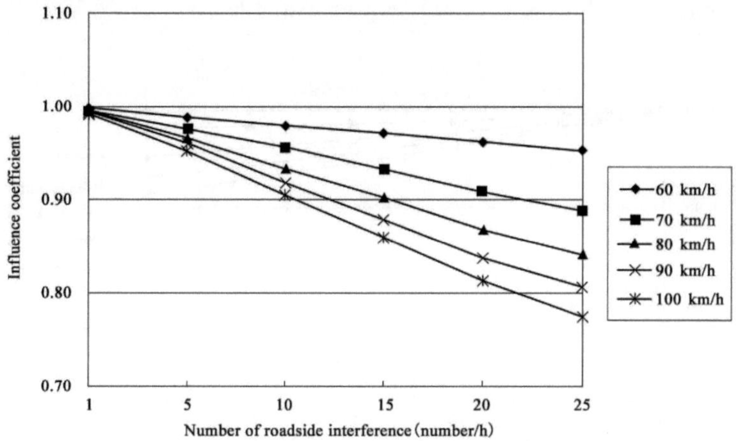


**Fig. B.4.11-1** Curve for implications of roadside interference on operating speeds ($W$ = 1.5 m). *Note* 1.Lateral distance from roadside interference $W$ = (hard shoulder + soft shoulder)/2 + 1.0 (m), taking a multiple of 0.5.2. When the operating speed of vehicles entering the road section with roadside interference, is less than 60 km/h or $W$ is greater than 2.5 m, the effect of roadside interference can be omitted. 3. Number of roadside interference = number of pedestrians + number of bicycles/3 + number of motorcycles/12 + number of parking vehicles × 1.25 (number/h)

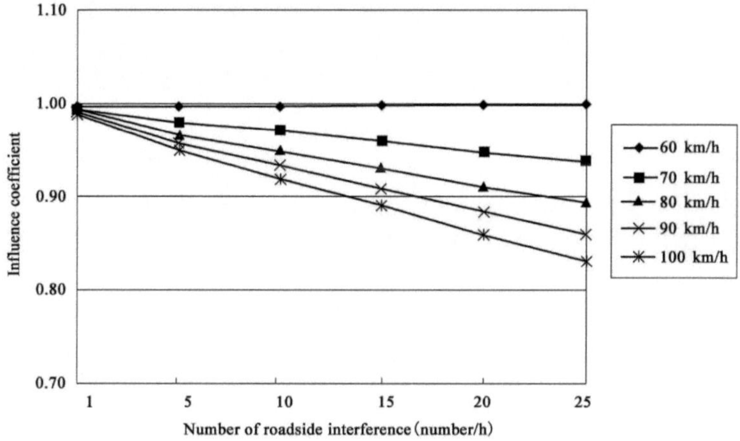


**Fig. B.4.11-2** Curve for implications of roadside interference on operating speeds ($W$ = 2.0 m). *Note* It is the same as Fig. B.4.11-1

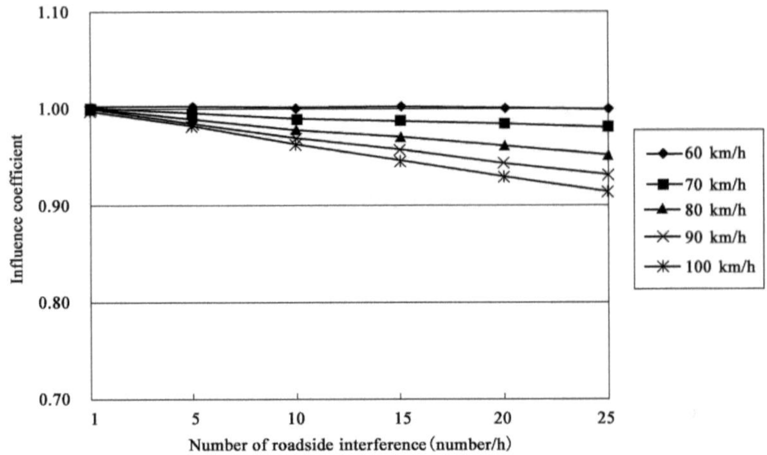


**Fig. B.4.11-3** Curve for implications of roadside interference on operating speeds ($W = 2.5$ m). *Note* It is the same as Fig. B.4.11-1

# Appendix C
# Calculation Methods for Clear Zone Width

C.0.1 Without considering the deployment of roadside protection facilities, clear zone width may be classified as suggested clear zone width and the effective recovery area width.

C.0.2 Suggested clear zone width shall be determined based on the conditions of the highway horizontal alignment indicators, sub-grade filling and cutting and operating speeds in compliance with the provisions as follows:

1. Suggested clear zone width should be determined based on the conditions of sub-grade filling and cutting for tangent sections obtained in Figs. C.0.2-1 and C.0.2-2.
2. Suggested clear zone width on curves should be modified based on the suggested clear zone width of tangent sections with the same sub-grade type and multiplied by the adjustment coefficient $F_c$, which can be obtained in Fig. C.0.2-3.

C.0.3 The effective recovery area width shall be the available width of a gentle, and unobstructed area extending outward from the outer edge of the traveled way, including paved shoulders, soft shoulders and utilizable fore-slopes in compliance with the following provisions:

1. If the fore-slope is flatter than 1:6, the effective width is the entire fore-slope width.
2. If the fore-slope is between 1:3.5 and 1:6, the effective width is 1/2 of the entire fore-slope width.
3. If the fore-slope is steeper than 1:3.5, the width of fore-slope is not an effective width because an errant vehicle cannot be driven through such a slope.
4. If there are masonry ditches or drainage channels without cover plates along the roadside, these do not form part of the effective width.
5. If there are non-removable street trees, flower planters, sign supports or other obstacles on the roadside, the width occupied by these obstacles does not form part of the effective width.

 
H. Wang, *Specifications for Highway Safety Audit*, JTG Standards and Specifications,
https://doi.org/10.1007/978-981-92-3132-4

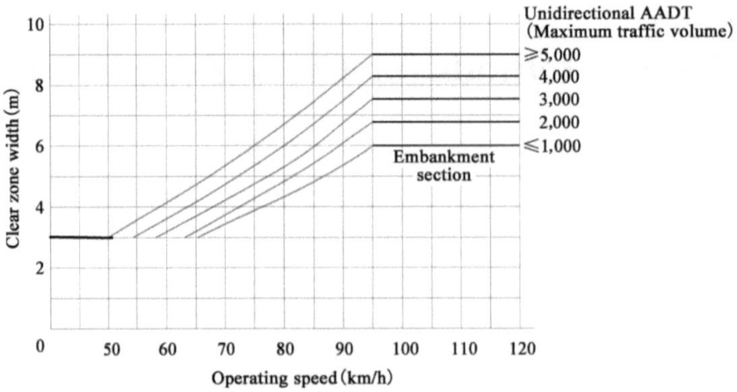


**Fig. C.0.2-1** Suggested clear zone width of tangent sections on embankments

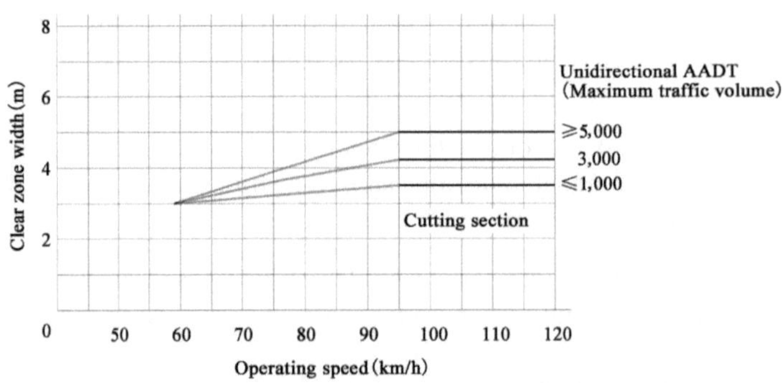


**Fig. C.0.2-2** Suggested clear zone width of tangent section in cutting

**Fig. C.0.2-3** Adjustment coefficient $F_c$ for suggested clear zone width on curves

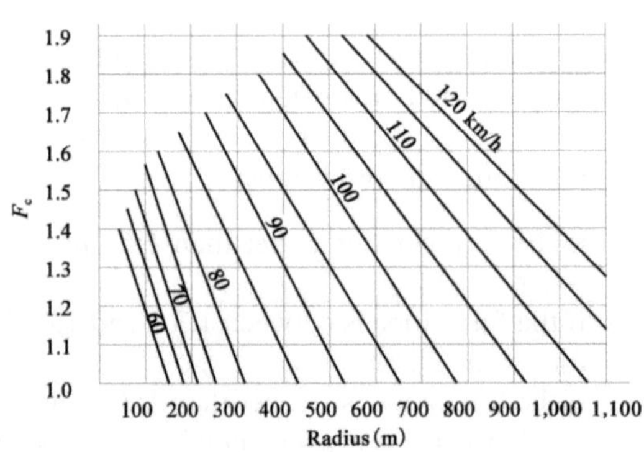

# Wording Explanation for the *Specifications*

1. The strictness in execution of the *Specifications* is expressed by using the wording as follows:

    (1) MUST—A very restrict requirement in any circumstances.
    (2) SHALL—A mandatory requirement in normal circumstances.
    (3) SHOULD—An advisory requirement.
    (4) MAY—A permissive condition. No requirement is intended.

2. Expressions used for reference to standards are explained as follows:

The standards for which a year is added to the standard number shall be the specific versions to be used. Otherwise they shall be the latest available versions.



H. Wang, *Specifications for Highway Safety Audit*, JTG Standards and Specifications,
https://doi.org/10.1007/978-981-92-3132-4

# Background to Provisions

## 1 General Provisions

1.0.1 The specification is formulated to unify the technical requirements and scope of highway safety audit and to ensure the quality of such audit.

1.0.2–1.0.3 The *Guidelines for Safety Audit of Highway* (JTG/T B05-2004) (hereinafter referred to as 'the *Guidelines*') are mainly applicable to the safety audit of motorways and Class-1 highways. Since the official promulgation and implementation of the *Guidelines* in 2004, the highway safety audit (road safety audit) has been carried out in China for more than 15 years. According to the results of questionnaire surveys in different provinces (municipalities directly under the central government), the concept of road safety audit has been widely accepted and applied in the construction and management process of highways at various technical levels. The demand for road safety audit is on the rise. Therefore, with the results of the questionnaire surveys and the current situation of traffic safety, adjustment has been made to the applicable scope of Clauses 1.0.2 and 1.0.3 in the *Specifications.* After the adjustment, the scope of highways' applicable for road safety audit is expanded from motorways and Class-1 highways to also cover Class-2 and Class-3 highways.

The applicable stages of road safety audit are adjusted from the stages of feasibility, design and operation to the stages of feasibility, preliminary design, detailed design, pre-opening and operation. The relevant contents of the post-audit stage are incorporated into the seventh chapter of operation stage in the *Specifications.*

The 'safety' noted in the specification refers to providing road users with highways, related facilities and a traffic environment conducive to the safety of all road users under the preconditions of the quality provision of highway structures and material in compliance with the national technical acceptance standards.

1.0.4 The dimension, payload mass and dynamic performance of design vehicles on highways serves as a main basis for determining the geometrical parameters of highways. The design vehicles are used as the representative vehicles for the road safety audit.



H. Wang, *Specifications for Highway Safety Audit*, JTG Standards and Specifications,
https://doi.org/10.1007/978-981-92-3132-4

At the preliminary and detailed design stages, the implications of higher proportions of vehicles within the forecasted traffic composition on highway geometric design shall be considered. If the proportion of large vehicles within the traffic composition is high, the use of operating speed of large vehicles is recommended for sight distance evaluation; for continuous upgrade sections of small-radius circular curve with super-elevation, small vehicles with higher operating speed while large vehicles with low speed and high center of gravity need to be considered.

At the operation stage (including post-audit or post-opening stages), traffic composition and traffic crashes need to be considered. Since the crash rate is high for large vehicles on sections of continuous downgrades, the consideration shall focus on speed control and braking-failure of large vehicles.

# 2 Terms and Definitions

2.0.1 Road Safety Audit (RSA) has become an established international road safety procedure. RSA is also called the highway safety audit, traffic safety audit and driving safety audit in China. Sometimes it is referred to as safety audit, safety assessment, safety evaluation or road safety assessment. The core of RSA is to evaluate the implications of highways, their facilities and traffic environment on road user safety. The purpose is to provide favorable conditions for traffic safety through audit, thus reducing traffic crashes, casualties and damage caused by traffic crashes and improving the safety level of the road network.

The "traffic safety" in this *Specifications* corresponds to the terms "driving safety" and "operational safety" in the current national and industry standards. In order to convey a more scientific and accurate meaning of words and to comply with common application, the related words in the Specification are collectively referred to as "traffic safety".

RSA originated in the United Kingdom (UK) in the early 1980s. RSA practices are now common in Australia, New Zealand, Denmark, Netherlands, the United States and other countries. RSA is defined as a formal, systematic, independent evaluation of the potential road safety problems associated with a new road scheme or road improvement scheme. The former Ministry of Communication (MOC) in China issued the Guidelines on November 1, 2004, as a recommended standard of road safety audit in the transportation industry. Meanwhile the concept of road safety audit has been officially promoted in China. According to international practices and the experience of road safety audit in China, for an objective and fair conclusion, the road safety audit project is usually undertaken by the third-party commissioned by the owner's agency or management departments. The third-party usually refers to an organization with the qualification of highway industry expertise, without any direct economic relationship with the design contractor, construction contractor, supervision contractor of the audited project, etc.

A road safety audit should be conducted by a multi-disciplinary team from the third-party. The RSA team should have appropriate experience in road safety engineering, crash investigation and traffic management. The team leader should be a



H. Wang, *Specifications for Highway Safety Audit*, JTG Standards and Specifications,
https://doi.org/10.1007/978-981-92-3132-4

specialist in RSA and the multidisciplinary team should consist of no less than three engineers.

The *Specifications* would introduce RSA technique requirements in detail. But the *Specifications* would not introduced the RSA process and management requirements in details. Because RSA has been an formal road safety process in China since 2004, most auditors are familiar with the process of RSA, which is similar to RSA in UK, USA, Australia, New Zealand etc. And the process of RSA is provided in *Application Handbook of Specifications for Highway Safety Audit* (Chinese version). In addition, management requirements should comply with local authority policies related to RSA.

2.0.3 Safety audit checklists are a popular prompt in RSA guidelines of UK, USA, Australia etc. The reference checklist and checklist formats are also provided in *Application Handbook of Specifications for Highway Safety Audit* (Chinese version), RSA auditors should complete and formulate complementary checklists incorporated with their audit experience.

# 3 Feasibility Stage

3.1 General

3.1.1 At this stage, the safety recommendations are provided of the factors that affect the traffic safety in each corridor proposal of a highway project covering the proposals for the engineering feasibility study and subsequent preliminary and detail design. At the same time, it is necessary to give consideration to the effects on the social and natural environment.

3.3 Audit Scope

3.3.2 If the design standard of the project is different from that of other projects connected via the same road network, the implications of the connection design and traffic management plan, etc. on traffic safety at the interface with the other highways shall be assessed.

3.3.3 The capacity of emergency response and evacuation after natural disasters or traffic crashes have been weak for a long time due to the disconnected highway network in China. Since the implementation of Emergency Response Law of the People's Republic of China in 2007, the Chinese government has gradually valued the evacuation emergency and rescue (EER). The provision aims to pay more attention to emergency rescue of traffic safety, etc. The safety audit is conducted to analyze the engineering proposals from the opinion of road network and to weigh whether the new highway could form a passage in the road system in case of disaster response and rescue. To layout extra-long tunnels and super-large bridges, etc. the route schemes are required to consider the difficulty of rescue and the feasibility of setting passage or lanes for emergency rescue.



H. Wang, *Specifications for Highway Safety Audit*, JTG Standards and Specifications,
https://doi.org/10.1007/978-981-92-3132-4

# 4 Preliminary Design Stage

4.1 General

4.1.1 At this stage, route proposals are audited to provide comparison and optimization of alternatives only from the view point of traffic safety.

4.4 Alternatives Audit

4.4.1–4.4.4 Comparison of alternative design proposals is an important part of the preliminary design stage. The recommended proposal often derives from the comparison of proposals by comprehensively considering project investment, environment, and other factors. By comparing and evaluating those factors related to traffic safety in alternatives based on the same level of details, audit results of each proposal are provided with qualitative descriptions, and the proposal is proposed from the perspective of traffic safety as a priority.

4.5 Audit on Design Elements

4.5.1 1 Design consistency refers to the highway geometry's conformance with driver expectancy. Operating speed-based design consistency is an important method of evaluation in the Guidelines and has been widely applied in RSA. From the special research program entitled Technical Research on the Operating Speed System, *Safety Audit and Engineering Application of Highways* (2007–2010), sponsored by the China MOT, a large amount of observed data of on-site speeds on typical highways have been subject to statistical analysis. The operating speed models were established, verified and improved. The theory and standards of RSA in China were enriched accordingly.

The *Specification* introduces the results of the above research and updates the evaluation criteria for the design consistency based on operating-speed and the prediction method of operating speed in the Guidelines. For highways with design speed at or below 80 km/h, road sections with constrained horizontal and vertical alignment features may possibly lead to the poor design consistency. For highways with design speeds of 100 or 120 km/h, the alignment criteria are generally suitable and the design consistency is good or fair under normal conditions. Therefore, motorways,



H. Wang, *Specifications for Highway Safety Audit*, JTG Standards and Specifications,
https://doi.org/10.1007/978-981-92-3132-4

Class-1, Class-2 and Class-3 highways with a design speed of 80 km/h and below is subject to evaluation of the operating speed-based design consistency in this Clause. The evaluation of operating speed-based design consistency include evaluating Operating Speed Consistency (OSC) between successive geometric feature segments, and the Speed Consistency (SC) between operating speed and design speed on the same segments.

3. The OSC is evaluated by an absolute value of operating speed differential between the operating speeds at the start and end points of the analysis unit. The method of dividing the analysis units is provided in the Appendix B of the *Specifications.* The operating speed gradient represents values of changes to operating speeds at the road segments of certain length (100 m is used in the *Specifications*), which reflects the sensitivity degree of changes to the operating speed within a certain length and is used as a supplementary parameter of the design consistency based on operating speeds on successive segments.

By considering indicators of the operating speed gradient, it could reflect the actual abrupt deceleration, and is helpful to screen road sections detrimental to the traffic safety. According to the experience of evaluating operating speed-based design consistency, if the operating speeds are increasing between the two successive elements at more than 20 km/h, there is generally little effect on safety. If the operating speeds at successive road sections is decreasing abruptly within a short distance, it is generally believed that the excessive deceleration brings potential risks and needs to be analyzed.

The absolute value of operating speed gradient $|\Delta I_v|$ is calculated with the Eq. (4-1).

$$| \Delta I_{\mathrm{v}} | = \frac{| \Delta v_{85} |}{L} \times 100 \tag{4-1}$$

Where:

| | |
|---|---|
| $\lvert\Delta I_v\rvert$ | Absolute value of operating speed gradient (km/ (h·m)); |
| $\lvert\Delta v_{85}\rvert$ | Absolute value of differences between the operating speeds at the start and end point of analysis unit (km/h); |
| $L$ | Length of the road section of analysis unit (m). |

The same road sections refer to the sections with same design speeds and cross sections.

4.5.2 1 (1) The drivers usually do not drive at the design speed on the highway, but at the appropriate driving speed according to alignment features, traffic situations and their own driving skills, etc. Compared with the design speed, operating speed reflects the driving expectation of most drivers and is more in line with the actual operational situation. The selection of the circular curve radius shall not only suit the changes to the geological and terrain condition of highways, but also be consistent with the upstream and downstream alignment features. If a small-radius circular curve has a different driving expectation, it may lead to sudden changes in operating speed and

**Table 4-1** Table for the relationship between the operating speed and the coefficient of friction

| Operating speed $v_{85}$ (km/h) | 120 | 100 | 80 | 60 | 40 | 30 | 20 |
|---|---|---|---|---|---|---|---|
| The coefficient of friction $\mu$ | 0.10 | 0.12 | 0.13 | 0.15 | 0.15 | 0.16 | 0.17 |

dangerous driving behaviors. Therefore, operating speed is adopted to evaluate the circular curve close to the minimum radius in the *Specifications*.

If the operating speed is used, the radius of a circular curve is calculated using the Eq. (4-2).

$$R = \frac{v_{85}^2}{127\,(\mu + i)} \tag{4-2}$$

Where:

$R$ Radius of the circular curve required by the operating speed at the road section (m);
$v_{85}$ Operating speed (km/h);
$\mu$ The coefficient of friction;
$i$ Grade of crown slope (%).

The coefficient of friction, $\mu$, need to take into account the friction coefficient between the pavement and tires as well as the comfort or discomfort of drivers and passengers. After weighing the above two factors, Table 4-1 lists the relationship between different operating speeds and the coefficient of friction.

(2) A clothoid is close to the trajectory of vehicles driving on the curve. The parameters and length of such clothoid directly affects the safety and visual effects of the alignment design, etc. The rational length of clothoids enables drivers unhurriedly to control their steering wheels. The appropriate changing rate of centrifugal acceleration can help drivers and passengers feel safe and comfortable. As it is common that the design speed is lower than the operating speed, the clothoid parameters decided by design speed often fail to meet the requirements of operating speed. Therefore, verifying the parameters and length of clothoids by adopting operating speed is required.

The minimum tangent length between curves is determined to ensure visual continuity of alignment. If the length of tangent between reverse curves is so short that the drivers have to turn frequently, drivers have difficulty in maintaining stable driving trajectory under speedy and stressful condition. If the length of a tangent between same-direction curves is too short, an error-illusion of reverse bending of two curves will lead to lack of continuity for the entire alignment. For determining the maximum length of a tangent, priority is given to the driver's visual effects and psychological endurance, thus avoiding driver's fatigue or over speeding behavior in driving. Evaluating the length of tangent between curves by using operating speed is more closely related to the actual conditions of driving, which is conducive to traffic safety.

2 (1) The sight distance of highways is an important factor to ensure traffic safety. The motorways and Class-1 highways are generally provided with central dividing

strips for dividing driving directions, thus avoiding the possibility of vehicles passing and overtaking in the opposing lanes. Therefore, only the stopping sight distance is determined for the motorways and Class-1 highways. While vehicles use the opposite lane for passing other vehicles on the Class-2 and Class-3 highways, determination of the stopping sight distance, meeting sight distance and passing sight distances shall be carried out for the Class-2 and Class-3 highways.

(2) According to the research results of the Special Report on Stopping Sight Distance of Trucks on Highways sponsored by the China MOC, large vehicles generally need longer stopping distances than passenger vehicles at the same design speed. Stopping sight distance of passenger cars is adequate on some road sections, but may be insufficient for trucks. Therefore, stopping sight distance of trucks is to be evaluated for relevant road sections with a large proportion of large vehicles on motorways and Class-1, the Class-2 and Class-3 highways. The relevant road sections generally include the accident-prone curves and exit ramps, sections with reduced number of lanes, as well as intersections or interchanges adjacent to the bridge pier, and at intersections close to crest vertical curves, etc.

(3) The operating speed reflects the actual driving speed. Therefore, the *Specification* recommends substituting operating speed for design speed to evaluate the stopping sight distance, meeting sight distance and passing sight distances. As the predicated operating speed is generally higher than the design speed, the available sight distance may not be sufficient or poor due to terrain conditions on some road sections. Therefore, the evaluation need to be analyzed for available safety measures.

3 (1) The safety analysis shows that continuous upgrades or downgrades (uphill or downhill) have a certain impact on the traffic safety. According to Clause 4.0.20 of the *Technical Standards for Highway Engineering* (JTG B01-2014) and the relevant background to their provision, the average longitudinal grade of the continuous upgrades or downgrades sections of the Class-2 (and below Class-2) highways in mountains shall be limited; for motorways and Class-1 highways, the rational average longitudinal grade shall be used to enhance the traffic capacity and operational safety. In the *Specifications* the grade, length and average longitudinal grade are to be evaluated for continuous upgrades or downgrades sections according to operating speeds and speed changes, speed variation, speed differential of vehicle types, and braking performance of large vehicles. It provides support for optimizing alignment, improving safety devices, and implementing traffic management and other comprehensive measures.

(2) The values of minimum radius for crest vertical curves are related to the requirements for mitigating acts of gravitational and vertical centripetal forces and sight distance. The values of minimum radius for sag vertical curves are related to the requirements for mitigating acts, headlight sight distance and the sight distance at under-crossings. The *Specifications* recommends that the operating speeds be used to the check the requirements for sight distance of crest vertical curves, mitigating acts of the sag vertical curves and the sight distance at under-crossings, etc. The length of the vertical curve is generally no shorter than the travelled distance of 3 seconds at the operating speed.

4 (2) When driving on the continuous upgrades, large vehicle speeds are sensitive to criteria such as grade and length of longitudinal grade, etc., and drivers tend to reduce speed within the process of continuously climbing. The indictors such as grade and length of longitudinal grade have little effect on the operating speeds of passenger cars. Thus speed differences between passenger cars and trucks increases, and result in higher frequency of overtaking and less safety. If large vehicles are climbing upgrades at a low speed, they will occupy carriageways for a longer time, and so have an adverse impact on the normal operation for other traveling vehicles. According to a special research on statistics data in US, when the truck speed is below the average speed by 16 km/h, the crash rate will increase significantly; when the truck speed is 24 km/h below the average speed, the crash rate will increase by 2.4 times. In order to alleviate the impact of continuous ascending sections (upgrade) on the traffic safety, Clause 4.0.8 of the *Technical Standards for Highway Engineering* (JTG B01-2014) stipulates that for motorways, Class-1 and 2 highways, climbing lanes shall be placed on long and continuous ascending grades if the highway capacity and operational safety would be affected. The width of a climbing lane shall not be narrower than 3.50 m. No climbing lanes may be needed for motorways with more than 6 lanes.

Therefore, the *Specification* stipulates that the necessity and location of climbing lanes need to be considered according to the forecast traffic volume, composition, service level, operating speed and other factors.

(3) The emergency escape ramp is a lane designed to enable vehicles with failed brakes to leave the carriageway lane, to slow down and to conduct the self-rescue. The Clause 4.0.9 of *Technical Standards for Highway Engineering* (JTG B01-2014) stipulates that for a long and steep descending grade, emergency escape ramps shall be placed in accordance with traffic safety evaluations. This is required to consider the necessity, locations and quantity of emergency escape ramps.

The siting of emergency escape ramps is determined by the simulation calculation of the brake temperature of a truck when going downhill. According to a special research and verification in China, the temperature of vehicle brakes exceeding 260 °C is used as a threshold for setting the emergency escape ramps. GSRS (Grade Severity Rating System) developed by the Federal Highway Administration (FHWA) uses the same value as the threshold. When the locations of emergency escape ramps are evaluated, the following points need to be considered:

The emergency escape ramps need to keep a proper departure angle with the main line. The places where small-radius curves approaches the end of continuous long downgrades are always prone to crashes. Generally, an emergency escape lane is placed along a tangent of the curve to ensure the smooth entry for the out-of-control vehicles. With respect to the probability of brake failures and psychology of drivers of the out-of-control vehicles, the site selection for an emergency escape ramp at the downstream end of continuous downgrades can enhance the utilization rate. The emergency escape ramps cannot be placed close to a densely populated area roadside. The alignment of emergency escape ramps need to be at a tangent to the mainline curve.

4.5.3 1 The width of the roadside clear zone consists of the suggested clear zone and actual widths of clear zone. The suggested clear zone refers to the minimum width required for the out-of-control vehicle returning to the pavement. The suggested clear zone is calculated based on considering the operating speeds, traffic volume at the road sections, subgrade filling and cutting and horizontal alignment features. The suggested clear zone in the Appendix C consists of the calculated widths of tangent and curved sections. The actual width of clear zone refers to the width of roadside which can stop errant vehicles or slow them enough to return to the roadway safely under the actual condition, considering the composition of subgrade section, side slope grade, forms of side drainage and roadside objects, etc., without considering the premise of protection facilities. Due to restrictions on land use and other reasons in China, the actual clear zone width may possibly be unable to meet the requirements of suggested clear zone width. When the conditions are not available, this Clause does not require that the actual clear zone width meets the requirement for the suggested clear zone width. Instead, the focus of the evaluation is then to check and verify whether the roadside objects treatment is effective to reduce or eliminate undesirable impact.

2 The fixed objects in the roadside clear zone directly affect the safety of errant vehicles, including drainage structures, sign, signal, and luminaire supports, utility poles, trees and similar roadside features. Combined with the practical experience of AASHTO (American Association of State Highway and Transportation Officials) (2011 version of the Roadside Safety Design Guide) and China, the common countermeasures are specified as follows:

Removing obstacles in the roadside clear zone; redesign the fixed object so it can be safely traversed; relocate obstacles to where they are less likely to be hit by out-of-control vehicles; using the breakaway facilities or structures to reduce the severity of collision by out-of-control vehicles. When the above four kinds of measures could not be implemented, and if the potential severity of crashes caused by run-off roadway vehicles is more serious than the collision with highway barriers, the placement of highway barriers or impact attenuator is considered to be implemented. If the above options cannot be implemented due to constrained conditions, then sight guidance and warnings are required to be established for obstacles.

In order to minimize the crash severity for the out-of-control vehicles, auditors are required to select and employ the treatment measures according to the actual situation.

4.5.4 4 When bridges are crossing over the audited highway, their piers and abutments can sometimes obstruct a driver's sight, which may result in insufficient sight distances on curve sections. It is necessary to evaluate the impacts of the bridge piers and abutments on the sight line. During the evaluation, the sight distance is checked and calculated according to the operating speed and the distance between the bridge piers or abutments and the lanes. When a bridge is just crossing over or located on the small-radius sag curve of the audited highway, the sight line may probably be obstructed by the superstructure of such a bridge. The impact of the superstructure on the visual condition of the audited highways needs to be assessed. The sight distance need to be checked and calculated according to the operating speed of vehicles on the main line, radius of vertical curve, clearance at the base of the beam of bridge etc.

4.5.5 2 The ponding and freezing of water close to the tunnel portals, etc. directly affects traffic safety. Where the tunnel portal is close to vertical curves, especially the crown of a vertical curve, almost a gentle slope, poor drainage often occurs.

4.5.6 2 The spacing between interchanges is an important indicator of traffic safety. The minimum spacing between interchanges and between an interchange and a service area, as well as the minimum spacing between adjacent entrances and exits etc. are presumed to meet the requirements for placement of traffic signs and the minimum distance required for maintaining a stable traffic flow. When the distance fails to meet the requirements for the general minimum distance, the rate of traffic crashes are increased due to frequent traffic diverging and merging. When the distance between the adjacent interchanges exceeds the spacing required for placing signs of exit warning, it has little effect on the safety. Clause 9.2.4 of the *Technical Standards for Highway Engineering* (JTG B01-2014) stipulates the spacing between successive interchanges. Clause 5.4.3 of the *Guidelines for Design of Highway Grade-Separated Intersections* (JTG/T D21-2014) stipulates the minimum spacing between the interchanges and other facilities (such as service and rest areas, etc.). In application, due to the restrictions of terrain condition, road network structure or other special circumstances, where the minimum spacing between the interchanges and other facilities fail to meet requirements, the provisions in the Clause 5.4.3 of the *Guidelines for Design of Highway Grade-Separated Intersections* (JTG/T D21-2014) is implemented. Clause 5.4.4 stipulates the requirements for the minimum net distance and principle of application if the distance is less than the net distance. The minimum net distance is determined by the distance required for the entire process of vehicles driving away from the mainline, including identifying and reading traffic signs, making decisions, finding the gap, changing lanes and confirming exits. To evaluate the minimum net distance, parameters such as operating speed, the number of lanes changing from inside lane to an exit need to be provided.

For determining the spacing between an interchange and a tunnel, the following points are supposed to be taken into account of:

(1) The spacing between the tunnel exit portal and the exit of mainline ahead: The auditor mainly considers requirements for placing indispensable signs (to provide regulations, warnings, and guidance information); in the case where the distance between the exit of the tunnel and the mainline exit ahead is relatively short in the mountainous area, the minimum net distance is provided. In Clause 5.4.5 of the *Guidelines for Design of Highway Grade-Separated intersections* (JTG/T D21-2014) the minimum net distance refers to the traveling distance in the process: Light adaptation, looking for gaps, changing lanes and identifying exits, etc. Since the minimum net distance is based on the assumption that drivers have identified and read the warning signs of exits and made decisions, etc. before driving out of a tunnel. Therefore, the auditors need to consider whether the precondition for the minimum net distance is established under the appropriate operating speed.

(2) The net distance between the mainline entrance and the tunnel portal ahead: the auditors require to consider factors such as looking for gaps, adjusting speed and

position of vehicles driving into the mainline, changes in the light environment at the portal section of tunnels, and the adaptation process of drivers between light and dark.

The distance between the interchange and the mainline toll station mainly considers the following:

(1) The distance between the toll station and the exit of the mainline ahead: the auditor mainly considers the need to meet requirements of location for the placement of indispensable advance signs (to provide regulations, warnings, and guidance information). Clause 5.4.6 of the *Guidelines for Design of Highway Grade-Separated intersections* (JTG/T D21–2014) stipulates the minimum net distance. The minimum net distance refers to the driving distance to complete the process of reading and recognizing signs, making decision, finding gaps, changing lanes and confirming exits, etc.

(2) The spacing between the toll station and the exit of the mainline ahead: the minimum net distance refer to the distance needed for drivers to read and comprehend the sign, make responses and decisions, look for gaps and adjust the vehicle, change lane and confirm the exit. If the minimum net distance is not enough to meet the basic requirements, an auxiliary lane is suggested to be designed. It is necessary to consider the implications of the placement of speed-changing sections and possible impacts of interweaving on traffic safety.

3 If types of entrances or exits of interchanges change frequently, drivers are prone to hesitate, make the wrong judgment or miss exits, and even run the risk of a crash. Section 5.6 of the *Guidelines for Design of Highway Grade-Separated Intersection* (JTG/T D21–2014) provides the requirements for consistency of exit types. The evaluation of exit types is required to consider the following:

The layout of gore noses are located uniformly preceding or after the crossing points (preferably be placed preceding the crossing points and single exit if deployed). When there is a clear distinction between the major volume and minor volume, the minor traffic flow ought to diverge at the right-side of the main traffic flow, rather than diverge at the left and right sides in turn.

5 If there is a distinct difference of service levels at road sections of interchanges or at the connecting highways, this would lead to serious turbulence of traffic flow and traffic bottlenecks. Operating efficiency is lowered and the safety level of traffic is reduced as well. Therefore, this item requires consideration of the traffic capacity and service level. The focus of the evaluation is on the interweaving sections where volumes change rapidly. Considering it is difficult to study the relationship between the service level and traffic safety in the preliminary design stage, 'may' is used in this Article.

4.5.7 1 This Clause applies to the evaluation of at-grade intersections on the Class-1, Class-2 and Class-3 highways and at the interchange ramp terminal of motorways. The accesses ought to be evaluated according to the *Specifications* or comply with relevant requirements of *Technical Standards for Highway Engineering* (JTG B01–2014). Clause 9.1.5 of the *Technical Standards for Highway Engineering* (JTG B01–2014) provides the minimum spacing between intersections depending on the highway class and function. A special research in China shows that when the spacing between successive intersections is less and the quantity of intersections is more, the roadside interference is more serious and more adverse to traffic safety. Accordingly, traffic delays are more serious. Therefore, this item requires the evaluation of the spacing between intersections. The locations of intersection are supposed to be considered in the evaluation as well.

3 Clause 9.1.2 of the *Technical Standards for Highway Engineering* (JTG B01-2014) stipulates that the traffic management of at-grade junctions is classified into three models, namely main road priority, non-priority and signalized, which shall be selected in accordance with the highway function, technical classification and traffic volume of the intersecting highway. Wherein the main road priority model is divided by stopping and give way, controlled by stop sign or yield sign on the minor road respectively. A study shows that some road users in China are unfamiliar with the traffic control model and rules of right-of-way at intersections, leading to the chaotic traffic condition and traffic crashes. The number of conflict points and conflict areas vary for different models of traffic control. The selected types of intersections require main highways to have smooth traffic, reduced number of conflict points and smaller conflict area. Therefore, this item proposes to evaluate the form of intersections based on the model of traffic control.

In addition, consistent forms of intersections on the same highway are helpful in achieving driving adaptability. If forms of intersections change frequently, drivers need to make complex judgements of the traffic management and operation characteristics when approaching. It would add to the drivers' workload. Drivers need to deal with a relatively large amount of information at intersection with higher vehicle speed and heavy traffic volume on arterial highways. So frequently changing forms of intersections is detrimental to driving safety.

4 The sight triangle within the scope of intersections is a key factor for ensuring the design safety of intersection. Clause 9.1.7 of the *Technical Standards for Highway Engineering* (JTG B01-2014) emphasizes the need to check intersection sight distances on classified highways. If the sight line is obstructed, safety measures shall be put forward in the evaluation.

The visibility area of the intersection sight triangle is related to the traffic control model used. Different models of traffic control have different requirements for the visibility at intersections. With the research results of the domestic and overseas national and industry standards, the common situations are listed below:

(1) If the intersection is using the non-priority control model or the main road priority and give-way on minor road, sight triangle is checked by reference to the operating speed. The diagram of the sight triangle at these intersections is shown in Fig. 4-1. If there are objects blocking the visibility in the sight triangle (including buildings, road-side trees and other greening objects, traffic engineering facilities and roadside safety devices, etc.), the visibility conditions need to be improved or the obstacles need to be removed from the sight triangle.

(2) If stopping on the minor road is employed as the traffic control model, it is perfect that sight triangle has the stopping sight distance (SSD) of each approach shown in Fig. 4-1. In the case of constrained conditions that could not guarantee the sight triangle surrounded by SSD, it is necessary to ensure that a sight triangle is

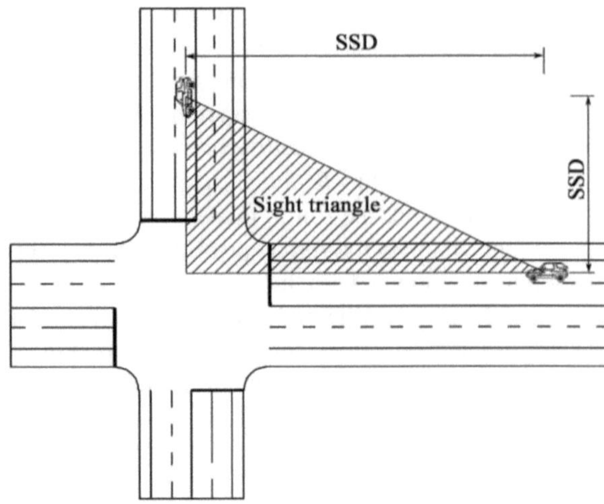


**Fig. 4-1** Sight distance triangle

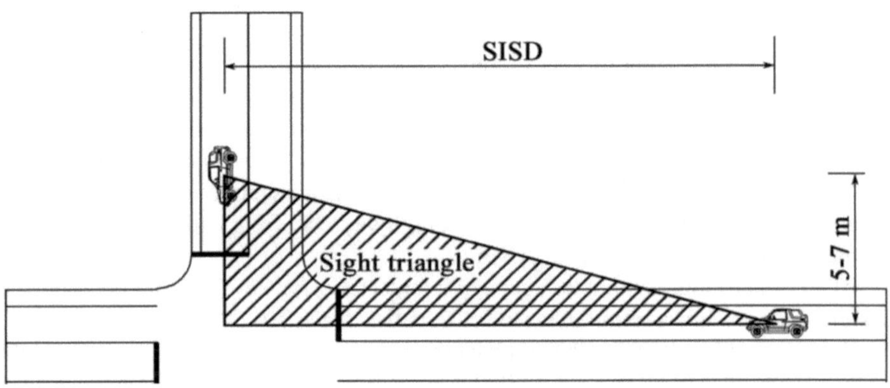


**Fig. 4-2** Sight triangle of the Safety Intersection Stop Distance (SISD)

provided by safety intersection stopping sight distance (SISD), which is composed of stopping sight distance on the main highway and a distance of 5–7 m from the minor highway to the central line to the main highway lanes, as shown in Fig. 4-2. Clause 10.3.2 of the *Design Specifications for Highway Alignment* (JTG D20-2017) stipulates that the SISD with different design speeds should be verified with the operating speed. SISD is calculated by using the Eq. (4-3), which was provided in the Guide to Traffic Engineering Practice of National Association of Australian State Road Authorities (NAASRA, Part 5) and verified in China.

$$\mathrm{SISD} = 0.833v_{85} + \mathrm{SSD} \tag{4-3}$$

Where:

SISD Safety Intersection Stop Distance (m);
SSD Stopping Sight Distance calculated using the operating speed (m);
$v_{85}$ Operating speed (km/h).

(3) If signalized control is employed as the traffic control model, according to the Policy on.

Geometric Design of Highways and Streets (the sixth edition) issued by American Association of State Highway and Transportation Officials (AASHTO), it is required that the first vehicle before the stopping line at each entrance lane should been seen by the driver of first vehicle on the entrance lane in the opposing direction. At the same time, it is necessary to check whether the left-turning vehicles have adequate sight distance and time for completing the left-turn. Research shows that it is an effective safety strategy to use signal control models for the intersection with high volumes of constrained sight distance, because a clear distribution of road rights of way, low vehicle speed and few traffic conflicts is provided under signal control model. According to the *Road Traffic Signal Lights* (GB 14887-2011) and the results of the report *Research on Traffic Safety Technology of Highway Intersections*, funded by the MOT, the evaluation of sight distance of signalized intersections requires the phase setting, signal timing, acceptable clearance, visible range of signal lights and etc. to be taken into account. For example, when vehicles can turn left and go straight within a phase, drivers on the main road ought to be able to see clearly a left-turning vehicle ahead and keep a safe distance. Drivers in opposing directions are also required to notice vehicles and provide enough distance to stop. If the pier and column, etc. at the median strip blocks the sight distance of left-turning vehicles, a phase to provide for the left-turning protection need to be considered, so that left-turning vehicles have adequate sight distance and time to complete the left-turn. If the right-turning vehicles are still allowed to travel within the red times, or flashing warning signal lights are employed (flashing yellow light on the main road and flashing red lights on the minor road) at night and during off-peak times, the sight triangle needs to be evaluated in accordance with the control model of stopping and give-way.

4.5.8 1 It is important to consider the comprehensive effect of traffic safety devices from systematic theory. Therefore, this Clause promotes design principles and types of traffic control devices and roadside facilities, etc. are required to be consistent with the main civil design, characteristics of traffic operation and requirements for safe operation. Hence assessment is conducted between various safety devices, traffic control devices and roadside facilities.

2 Service and rest areas can offer drivers, especially drivers of large vehicles, with a place for parking, relaxation, inspection and repairs. If conditions are available, facilities used for adding water and cooling for vehicles may be provided. Information such as the grade, length, dangerous points, emergency escape ramps for continuous downgrade sections can be indicated on the signs in the service and rest areas, etc., so that drivers of large vehicles are conscious of the dangers of brake failure and also provided with a suggested safe downhill speed.

Practice has proved that regular setting of service and rest areas is an effective way to improve the safety at the start of downhill sections and within continuous downhill sections if favorable sites with good visual condition are available.

If a service or rest area is located close to the end of continuous downgrade sections or preceded by a small-radius circular curve, or the location is significantly lower than the mainline, it would lead to rapid acceleration and deceleration for drivers, affecting traffic safety. In addition, if the layout of land for service and rest areas is irrational, vehicles are likely to be congested at the entrance and reduce the safety level. Therefore, this Clause stipulates the evaluation of the location, scale, alignment and design of service and rest areas.

For the combined placement of service areas and interchanges, the evaluation requires attention to simplify the complexity of traffic flow direction based on the continuity and convenience of movement. It is also necessary to verify the distance between adjacent exit or entrance ramps, and the distance from the ramp exit to the rest area to meet the requirements of Clause 12.2.7 of *Guidelines for Design of Highway Grade-Separated Intersections* (JTG/T D21-2014).

3. The operating speed and sight distance before arriving at a toll station is greatly influenced by alignment. If the toll station is located close to the end of a continuous long downgrade or ramp, or preceding a sharp curve, it is not conducive to slowing-down and safe entry into the toll station. Accordingly, this item is used to evaluate the position of toll stations. The evaluation does not only need to consider the technical requirements for location of toll station, alignment of toll station square provided in Clause 6.2.2 and Section 8.6 of *Guidelines for Design of Highway Grade-Separated Intersections* (JTG/T D21-2014), but it also needs to evaluate other features such as visibility conditions, operating speed transition and longitudinal grade influences on the operating speed of large vehicles.

5 Sections susceptible to adverse weather mainly refer to the sections which are always subject to severe impacts of heavy rain, snow, fog, ice, crosswind and snow-drifts, etc. The design of surveillance facilities still needs to consider the requirements of the Internet and sustainable development, to improve equipment efficiency and reduce wastes at the time of upgrading.

6 When the highway upgrading and re-construction project is based on existing road sections with high accident rates, serious traffic congestion and frequent weather hazards, the design programs shall be in accordance with the requirements of Clause 4.0.7 of *Guidelines for Design of Traffic Engineering and Facilities of Expressway Reconstruction and Extension* (JTG/T L80-2014). The evaluation is focused on comprehensive improvement measures for highway upgrading and reconstruction projects.

# 5 Detailed Design Stage

## 5.1 General

5.1.1 This stage aims to provide suggestions for optimizing the design of traffic control devices and roadside facilities by auditing the compatibility of traffic control devices and roadside facilities, alignment features, structures, traffic surrounding etc.

## 5.2 Methodology

5.2.2 Driving simulation is usually conducted in driving simulator systems. A driving simulator system is an experimental platform for testing the driving behavior. It consists of a driving simulator, a scene projector, computer and software system, etc. The driving simulation is not only used to create a three-dimensional scene simulation environment of terrain and geologic conditions, road alignment, traffic safety facilities and traffic flow characteristics, etc., but also to enable drivers to really feel the changes of engine sounds, vehicle vibration and turning centrifugal force in the process of simulated starting, acceleration, deceleration, turning and braking, etc. The method of driving simulation is used to provide dynamic tests and to record the vehicle speed, acceleration, driving trajectory, drivers' psychological and physical data of experiments, etc. It is an effective means to test the changes of the driving behavior and traffic safety under the complex condition of road traffic.

## 5.4 Audit on Design Elements

5.4.1 1 From the perspective of ensuring traffic safety, the first item of this Clause is required to evaluate the application of super-elevation values. The operating speed, large vehicles' proportion, weather conditions and other factors need to be considered for the evaluation of the super-elevation value and the maximum super-elevation value. Consideration is given to the actual condition of vehicle operation and the predicted operating speeds. If the super-elevation values obtained by using operating speed is greater than that by using design speed, it is necessary to compare the scheme for increasing the super-elevation value with the scheme of maintaining the original design or adjusting the speed control facilities.

The value of super-elevation is calculated based on the Eq. (5-1).



H. Wang, *Specifications for Highway Safety Audit*, JTG Standards and Specifications,
https://doi.org/10.1007/978-981-92-3132-4

$$i = \frac{v_{85}^2}{127R} - \mu \tag{5-1}$$

Where:

$i$ Super-elevation value;
$v_{85}$ Operating speed (km/h);
$R$ Radius of circular curve (m);
$\mu$ Lateral friction coefficient between the pavement and tires.

The super-elevation need to consider not only cars with higher speed but also large vehicles with lower speed. Using a reasonable super-elevation could effectively avoid overturning and slipping crashes. Centrifugal force increases with vehicle speed, so higher speed needs a larger super-elevation to balance the centrifugal force. However, if the speed is slow and the centrifugal force is small, large vehicles with low speed and higher centers of gravity run the risks of slipping along the pavement with a combination of super-elevation and longitudinal grades. In general, the slipping occurs before overturning. But if the large vehicles have a higher center of gravity after heavy loading, it is likely to overturn.

In addition, if a steep longitudinal gradient and sharp circular curve are combined, the operating stability of trailers and similar large vehicles will be adversely affected. When the semi-trailer brakes on a section of downgrade, the movement of cargo reduces the lateral force of the rear axle of the vehicle. If the operating speed of the vehicle is close to the maximum safe speed on the curve, the rear wheel may deviate from the trajectory, causing the vehicle to overturn. In order to offset the adverse effects on the stability of trailers, the super-elevation of circular curves located on steep gradients (generally more than 3%) needs to increase appropriately and be calculated according to the Eq. (5-2):

$$E_{\min} = E + \frac{i_{\text{longitudinal}} + E}{6} \tag{5-2}$$

Where:

$E_{\min}$ Minimum value of super-elevation of large-grade descending sections;
$i_{\text{longitudinal}}$ Absolute value of longitudinal grade;
$E$ Super-elevation designed on less than 3% grade.

2 According to investigations, while vehicles are driving on a circular curve, the radius of the rear wheel's trajectory radius is smaller than that of the front wheel. For large articulated trucks with containers, the risk of vehicles intruding into the outside of lane would increase. When the proportion of heavy vehicles is relatively high, by widening road sections of small radius circular curve on motorway ramps, Class-2 and Class-3 highways, the safety of articulated trucks while turning and passing can be ensured. *Design Specifications for Highway Alignment* (JTG D20-2017) provides three types of widening of circular curve sections with the corresponding values of widening and application conditions. Clause 9.4.1 and 9.4.2 of the *Guidelines*

*for Design of Highway Grade-Separated Intersections* (JTG/T D21-2014) stipulates the traffic conditions and widening values of the pavement on the ramp circular curve sections. According to the investigation, the track of vehicles usually offsets towards the internal side of the curve when entering small radius circle curves, and offsets towards the external side of the curve when exiting the curve. For small radius curves of the same radius, the faster a vehicle approaches the curve, the more the offset towards the internal side of the curve will be, then vehicles will pass through the center line of roads. When the operational speed of vehicles is too fast, the risk of vehicles crossing the center line still exists, even if widening has been made in accordance with the requirements of *Design Specifications for Highway Alignment* (JTG D20-2017). For the situation above, offsetting the center line towards the internal side of the curve or applying channelized design for center markings is recommended, in order to prevent drivers from occupying the opposite lane and to guide vehicles to drive along the safe trajectory.

4 According to the investigations, in some highway sections with climbing lanes, many large vehicles are driving on climbing lanes, leading to most of lanes being occupied by large vehicles and some small vehicles taking a risk to overtake by using climbing lanes as overtaking lanes. This is often associated with irrational provisions of signs and markings on climbing lanes. Therefore, it is necessary to evaluate signs and markings on the climbing lanes. For example, the markings of lanes before climbing lanes are directly connected with makings of the climbing lane, and warning sign ought to require large vehicles drive at the right side of road sections with climbing lanes.

5 In order to maximize the function of escape ramps under emergency situations, the following matters need to be considered during the evaluation of escape ramps:

(1) Approach:

Good visibility is needed on the approach to the entrance to the emergency escape ramps, so that drivers can clearly see the entire alignment of the emergency escape ramps. The intersecting line of the escape ramp approach with an arrester bed ought to be perpendicular to the driving direction of vehicles entering the escape ramp, to ensure a simultaneous entrance for both front wheels of vehicles, and to avoid the roll-over accidents at the entrance due to uneven friction between the wheels on both sides. According to engineering experience, and the method of direct-deceleration lane adopted for the escape ramp approach, the angle of departure for the ramp (between efflux angle and the main line) is supposed to be small, usually 5 degrees or less. Even if constrained by terrain, the maximum value should be no more than 10 degrees.

(2) Horizontal alignment:

A curved alignment ought to be avoided for the horizontal alignment of escape ramps. Curves can reduce the function of escape ramps with wheels turning sharply and unexpectedly with an out of-control vehicle would cause roll-over or collision with the barriers, sometimes more serious secondary accidents could be initiated.

(3) Vertical alignment:

A single ascending grade is recommended for the vertical alignment of gravity ramps (arrester bed). If multiple grade combinations are required due to the constraints of terrain conditions and engineering quantities, small-radius crest curve needs to be avoided, and the alignment of escape ramps needs to be visible to the drivers of out-of-control vehicles.

(4) Length and grade:

The length and grade of the emergency escape ramps are comprehensively determined by factors such as terrain conditions, engineering quantities, exit speed of out-of-control vehicles and materials of arrester bed, etc. The length of the escape ramps includes the length of the diverging section, vertical curve and arrester bed. The length of the arrester bed is the length that has the actual effect in deceleration. To reduce the length of escape ramps and cut down engineering quantities, an appropriate increase in upgrade slope of the escape ramps is usually considered. The length of escape ramps can be reduced by increased the upslope, however, a drivers' sense of fear would be initiated due to overly steep slopes so that they do not dare to drive into these escape ramps. In addition, vehicles parking on escape ramps are supposed to not slide due to the increase of upgrade slopes. Usually the grade of a ramp upslope is limited to 8%–10%.

(5) Materials of arrester beds:

Sufficient stopping distance is a prioritized factor for escape ramps. Materials for the arrester beds of escape ramps needs to be pure, not easily compacted and have a high coefficient of rolling resistance. The frequently used materials for arrester bed are sands, gravels, macadam and cobbles. The depth of aggregate pit for arrester bed is 50 to 100 cm typically. In order to reduce the compaction while vehicles are entering into the arrester bed, the depth of aggregate pit transits from 10 cm to the whole depth in the first 30 to 60 m of the arrester bed.

(6) Terminal processing:

To ensure the safety of vehicles and drivers, anti-collision walls and buffer facilities need to be installed at the end of escape ramps. Generally, aggregate piles, anti-collision sand barrels or waste tires, etc. are placed at the front of anti-collision walls. The effective braking distance of escape ramps cannot be replaced by anti-collision walls and buffer facilities, the speed of vehicles be reduced to 20 km/h or less before vehicles hit an anti-collision wall and buffer facility to ensure the safety of vehicles and drivers. Otherwise, if conditions are severely restricted, arresting nets are recommended for escape ramps.

(7) Drainage:

If water penetrates into the voids of the aggregate sand bed, it will lead to higher compaction for the sand bed and lower rolling resistance. Optimum design for drainage facilities is required to ensure the cleanness and braking effect of aggregates for arrester beds.

(8) Traffic safety and control facilities:

Highway safety devices such as signs, markings, barriers and ancillary facilities such as lighting need to be installed so that escape ramps would operate under emergency situations. When a number of emergency escape ramps are placed on the continuously long descending sections, the information of the number and location of each escape ramp needs to be included in the sign design.

6 (2) In accordance with Clause 3.0.7 of the *Guidelines for Design of Motorway Upgrading and reconstruction and Extension* (JTG/T L11-2014), after asymmetrical widening or single-side separating and extension, the existing two-way traffic is converted to one-way traffic, the belt-like divider for separating the same direction traffic flow is installed on the road sections where located at former medians. The former central dividing strips are converted into lane change-over zones for changing lane, where former medians are rebuilt as pavement.

Clause 5.1.3 provides that "the highway of reconstruction, upgrading and extension should be widened on both sides. Widening at one side of the subgrade may be adopted unless restricted conditions apply. If one side only is widened, the lane change-over zones on the original road and traffic engineering, etc. shall be strengthened." The widening on one side is usually used due to restrictions of terrain conditions or interchanges and other large structures. Considering the demand for changing lane and the complicated operation conditions before the entrance and exit of interchanges, the location and length of lane change-over zones as well as traffic control devices needs to be evaluated to ensure safety.

5.4.2 1 The road sections with different pavement materials or skid-resistance susceptible to reduction mainly include continuous upgrades, continuous long and steep downgrades, tunnel portals, large bridges, bridge-tunnel sections, toll stations and road sections passing villages and towns. Large vehicles, especially overloaded vehicles, emit a lot of gas while climbing up the continuous upgrade section, coefficient of friction resistance will be reduced due to the deposit of gas onto road surface. For some continuous long and steep downgrade sections, where cooling and braking by dripping water is frequently used, the coefficient of friction resistance could also be reduced by water dropped onto the road surface, resulting in vehicles skidding and unstable driving.

3 Trapezoidal or rectangular side drains and gutters without grates are hazards to run-off-the-road vehicles, as they do not afford them an opportunity to return to carriageway. Traversable (such as shapes of a triangular and shallow dish, etc.) roadside drains and gutters are more favorable to enable out-of-control vehicles to return safely to the carriageway.

If the crown of a vertical curve is almost a gentle slope, especial the sections of super-elevation transition close to the common tangent point of S-shaped horizontal curves, poor drainage would be potentially hazardous to traffic safety.

Except for considering drainage facilities design, the other measures such as using a relatively higher of super-elevation transition rate to reducing the length of gentle slope sections, or adding one or more crown lines can be considered to eliminate or alleviate the problem of ponding of water for the gentle slope section.

4 For an upgrading and reconstruction project, the catchment area of road surface is enlarged after road widening, it is necessary to evaluate whether drainage facilities are sufficient for the new drainage requirements. Clause 7.1.5 of *Guidelines for Design of Motorway Upgrading and Reconstruction and Extension* (JTG/T L11-2014) stipulates that the function of drainage facilities on the existing median strip and super-elevation sections shall be maintained or improved. Damaged drainage

facilities shall be repaired. The drainage facilities shall be rebuilt if the drain capacity of facilities would not be suitable for the upgrading and reconstruction project.

5.4.3 1 Once errant vehicles collide with piers, abutments of overpass bridges, it would cause damage to both vehicles and overpass bridges. If the piers and abutments of a bridge are located in the roadside clear zone, effective protection is to consider the collision-deformation and requirement of minimizing damage to both the vehicle and bridge structure.

4 The main safety issues caused by the crosswind on the bridge deck include the heeling, side-slipping and latero-deviation of vehicles. The heeling refers to the issue that a vehicle overturns caused by strong crosswind; the side-slipping refers to the phenomenon of vehicles losing control under the combination of high wind speed and the wet and slippery pavement, the main reason is lateral friction force being less than the crosswind force; the latero-deviation refers to the phenomenon that vehicles intrude into adjacent lanes due to the excessive and cumulative lateral offset if drivers do not turn the steering wheel with impact of crosswind. The above three issues are related to pavement condition (friction coefficient), highway alignment (curves, longitudinal grades, horizontal slopes, etc.), vehicle models and loads, driving speed, as well as wind speed and direction, etc. The study shows that side-slipping occurs firstly under the same conditions of vehicle model and driving speed with the increase of crosswind. The risk of lateral sliding increases with the increase in driving speed and reduction of pavement adhesion coefficient. Therefore, the fourth item of this Clause requires the audit of the effects of crosswind on the traffic safety of bridge decks.

5.4.4 2 A relatively high accident rate appears at some tunnel portal sections. According to investigations, a vehicle is prone to side-slip or collide with portals if there is a large difference in pavement skid-resistance performance between the inside and outside of tunnels. Clause 8.0.1 of *Technical Standards for Highway Engineering* (JTG B01-2014) stipulates that the skid resistance shall be sufficient and consistent on the pavement within a tunnel and on the sections approaching the tunnel. In recent years, a special research showed that the connecting road sections inside and outside the highway tunnel generally adopt the same pavement materials, with the same consistent design of skid resistance. However, there can be a difference in the skid resistance of a connecting road section after a period of opening to traffic, due to conditions of topography, climate and vehicle usage, etc. The connecting road sections refer to the connected road sections adjacent to tunnel portals inside or outside tunnels with certain lengths.

Therefore, the second item of this Clause stipulates that the measurements for improving skid resistance at the tunnel portal ought to be evaluated.

An investigation in China shows that, for some highway tunnels, cement concrete pavement is adopted inside tunnels while asphalt concrete pavement is adopted outside tunnels and in the transitional section (about 100 to 300 m) inside tunnels.

At the interface of different pavement materials of transitional sections, due to reflective effects of lighting and lamps, drivers are likely to misjudge the pavement conditions ahead, resulting in hurry-wheeling, braking and other error operations.

Under wet or slippery pavement conditions, errant vehicles are likely to side slip and lose control.

Based on the above issues, the audit ought to consider the implication of transition positions, alignment, drainage, lighting on safety.

3 The lighting of highway tunnels, especially at the tunnel portals has a great impact on the driving safety of vehicles, which is reflected in the following two aspects:

On one hand, a transitional period of adaptation is required for a human's visual performance from light to dark. Because there is a distinct difference between the brightness and darkness inside and outside the tunnel, it is so dark that drivers approaching the portal usually could not identify hazards, which is known as the black-hole effect. Due to the black-hole effect, drivers cannot see the bright light but the black hole when they drive out of the tunnel portal, which makes it difficult for them to identify the alignment outside the tunnel portal or identify obstacles on the pavement. At that moment, drivers probably respond by sharp deceleration, braking, wheeling or other operations. Once the errant vehicle loses control, collisions with tunnel walls or rear-end would occur.

Although the situation at night is contrary to that during the day, the risk of collision is the same.

On the other hand, dazzling and strong glare appear during daytime near the tunnel exit portal, known as white-hole effects. The white-hole effects will reduce the view of drivers before they drive out of the tunnel portal and bring about errors in judgment and operation.

Therefore external and internal lighting of tunnel portals and operating speed needs to be evaluated. In order to ensure the driving adaptation, the evaluation is supposed to comply with the relevant provisions of *Specifications of Design Highway Tunnel Section II*: *Traffic Engineering and Affiliated Facilities* (JTG D70/2-2014) and *Guidelines for Design of lighting Highway Tunnels* (JTG D70/2-01-2014).

Tunnel ventilation is evaluated according to the relevant provisions of the *Guidelines for Design of ventilation of Highway Tunnel* (JTG D70/2-01-2014), with consideration of fog design concentrations affected by heavy vehicle proportion, operating speed and tunnel gradient. Ventilation indicators in the long tunnel and super-long tunnels, accumulation of pollutants under traffic congestion and fire disasters also need to be considered.

RSA conclusion should quote those documents related to traffic safety, including design review or evaluation for tunnel lighting ventilation or firefighting. The clause focuses on the assessment on lighting, ventilation, fire and surveillance facilities design, etc. under the normal conditions of traffic operation. Other abnormal conditions such as traffic congestion and fire disasters, special evaluation reports are suggested if necessary.

Given the complexity of the evaluation and the feasibility of related provisions, 'should' is used in the clause for the *Specifications.*

4 Median crossovers at tunnel portals and ancillary facilities are desirable to ensure traffic safety. Facilities of information release and lane guidance for long tunnels and

extra-long tunnels ought to be considered and be located before separated subgrades and changing lane for traffic flow adjustment situation.

While travelling in tunnel chains (group tunnels), vehicles will enter and leave portals frequently, driver's psychological and physical conditions would be affected by the variation of brightness and darkness. The segments with the tunnel chains are usually connected with bridges and tunnels, leading to great difficulties in EER. Clause 8.0.5 in the *Technical Standards of Highway Engineering* (JTG B01-2014) stipulates that successive tunnels, of which the distance between portals of any two tunnels is shorter than the distance travelled during 6 s at design speed, shall be designed in a systematic and integrated way by taking into account the ventilation, lighting, safety devices, and administration facilities as well as disaster relief, emergency aid and so forth. Therefore, the traffic safety tunnel chains and their ancillary facilities related to needs to be evaluated as a whole.

5 It will be common that a one-way traffic flow is diverged into two tunnels in highway upgrading and reconstruction projects. In order to facilitate the crash rescue, resolve safety issues caused by lane-selection of drivers and the unevenness of traffic volumes, Clause 10.1.6 in the *Guidelines for Design of Highway Upgrading and Reconstruction and Extension* (JTG/T L11-2014) stipulates that when new, expanded and excavated tunnels serve as a mutual escape passage for an existing tunnel, the placement of the crossing passage shall be in compliance with the requirement of newly built tunnels.

5.4.5 1 The control curve refers to a curve for which the alignment features are controlled according to design speed of basic ramp sections. Usually traffic safety at interchange sections is affected by a higher operating speed differential between the gore nose and the control curve.

If an appropriate length of the transition section is provided from the gore nose to the origin of ramp control curve, operating speed can be reduced continuously and proportionally.

In order to control the change of operating speed on the transition section within a reasonable range, the curvature radius of the horizontal curve at any point on the transition section needs to be compatible with the operating speed, which can avoid the safety issues arising from sharp steering or emergency braking.

Therefore, this Clause stipulates the evaluation of the length from the gore nose to the start of the alignment control curve of a ramp, and of the radius of circular curve on traffic safety on traffic safety. During the evaluation, the passing speed at the gore nose is important basic data for determining the alignment indicators of exit ramps design, which is determined by field observation or simulation of the deceleration process for operating speed of similar exit ramps. Clause 4.3.3 of the *Guidelines for Design of Highway Grade-Separated intersections* (JTG/T D21-2014) stipulates the speed of passing the diverging gore nose of exit ramps; and Clause 8.4.2 (JTG/T D21-2014) provides requirements for the design of transition section.

2 The interchange ramps are usually single-lane or two-lane highways, whose characteristics of operating speeds are similar to the Class-2 and Class-3 highways under the conditions without roadside interference. The evaluation focuses on ramps with constrained alignment indicators. Considering the applicability of operating

speed prediction models and their parameters, it is preferred to use self-calibrated operating speed prediction models by observing vehicle operation in the region where the projects are located.

3 The basic ramp road sections refer to the sections of the ramp where traffic operation is not affected by diverging, converging and weaving. Clause 4.4.5 of the *Guidelines for Design of Highway Grade-Separated intersections* (JTG/T D21-2014) requires that while there are obstructions or objects such as bridge piers, barriers, cutting slopes, plants, etc., and the radius of the circular curve is small, stopping sight distance shall be evaluated for the inside lane of the curve, and decision sight distance shall be evaluated for sections preceding the gore nose. If objects (such as barriers, glare screens, plants, etc.) block of view, and the radius of circular curve is small, stopping sight distance shall also be checked for the outside lane of the curve adjacent to the median. Clause 4.4.4 (JTG/T D21-2014) stipulates that on road sections where traffic composition is mainly large vehicles or the visibility of heavy vehicles is affected, the sight distance of intersecting roads and ramps shall be no less than the stopping sight distance of trucks. Clause 4.4.3 (JTG/T D21-2014) stipulates the basic requirements for the stopping sight distance on the basic road sections of ramps at different design speeds. Operating speed is used in the audit of ramps.

While evaluating the visibility conditions of diverging and merging noses, the stroke length corresponding to the operating speed is usually calculated according to Fig. 5-1, and the visibility of the enclosed sight triangle needs to be evaluated.

5.4.6 3 In considering the function, classification and traffic volumes of the intersecting highway, combined with the current national and industry standards *Specifications for Design Highway Alignment* (JTG D20–2017) and *Specifications for Setting and Installing Road Traffic Lights* (GB 14887-2011), three control modes ought be audited including main road priority, no-priority and signalized intersection.

5.4.7 1 Signs are set for travelers who are not familiar with the road network, but have certain plans for travel routes. The placements of multiple signs and the placement of successive traffic signs shall be coordinated to avoid sight blockage. While surveillance facilities, lighting facilities, information dissemination devices and structures exist on road sections with signs, positioning also needs to be coordinated to avoid mutual obscuring.

The information on the traffic signs need to be repeated or successively set for the transmission step by step. If the amount of information on the sign is appropriate, it is supposed to enable drivers operating vehicles at the normal speed to identify and understand the information. On the contrary, if the information on the sign is not accurate or the amount of information is excessive, it would affect the visual effects and traffic safety. Therefore, this Clause stipulates the audit of the rationality, the continuity, validity and quantity of information on the signs.

The dimensions of traffic signs and the font height are evaluated according to Tables 5-1, 5-2, 5-3 and 5-4. While setting signs for multi-lane highways, the case that the drivers' sight of vehicles on the inside lane may be blocked by large vehicles on the outside lane needs to be considered, additional signs above the inside lane or on a gantry-type support structure are generally required.

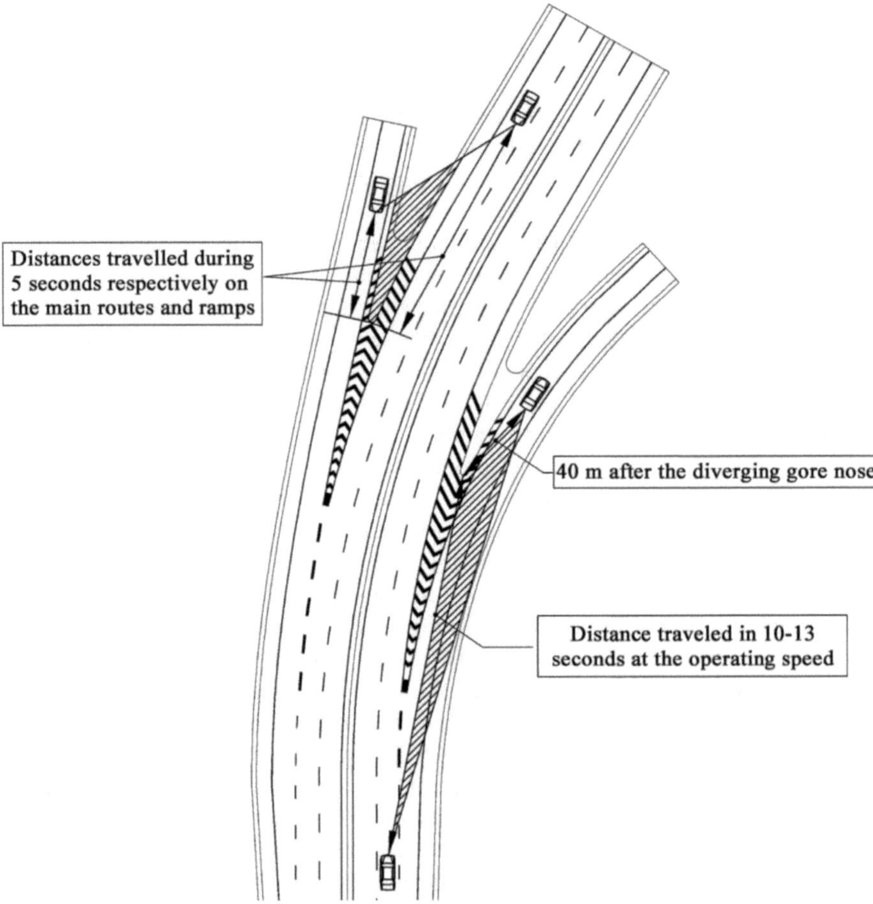


**Fig. 5-1** Figure of sight distance parameters of interchanges

**Table 5-1** Relationship between the dimension of warning signs and operating speeds

| Operating speed (km/h) | ≥100 | 90–70 | 60–40 | <30 |
|---|---|---|---|---|
| Length of triangle side (cm) | 130 | 110 | 90 | 70 |
| Width of black side (cm) | 9 | 7 | 6 | 5 |
| Radius of circular angle of black side (cm) | 6 | 5 | 4 | 3 |

4 When evaluating roadside barriers and median barrier, information on horizontal and vertical alignment, cut and fill, traffic volumes, composition and operating speed need to be collected, and the distribution of various obstacles within the clear zone as well as the intersections with other highways and railways need to be acquired. During the evaluation, it is necessary to determine whether to set barriers according to the severity of the crash caused by the vehicle run-out-of-road, and check the

**Table 5-2** Relationship between the dimension of regulatory signs and operating speeds

| Operating speed (km/h) | | ≥100 | 90–70 | 60–40 | <30 |
|---|---|---|---|---|---|
| Circular signs | External diameter of circular (cm) | 120 | 100 | 80 | 60 |
| | Width of red side (cm) | 12 | 10 | 8 | 6 |
| | Width of red line (cm) | 9 | 7.5 | 6 | 4.5 |
| Triangle signboard | Length of triangle side (cm) | – | – | 90 | 70 |
| | Width of red side (cm) | – | – | 9 | 7 |

**Table 5-3** Relationship between the size of indicative signs and operating speeds

| Operating speed (km/h) | ≥100 | 90–70 | 60–40 | <30 |
|---|---|---|---|---|
| Diameter of circular (cm) | 120 | 100 | 80 | 60 |
| Length of square side (cm) | 120 | 100 | 80 | 60 |
| Length of rectangular side (cm) | 190 × 140 | 160 × 120 | 140 × 100 | – |
| One-way street sign (rectangular) (cm) | 120 × 160 | 100 × 50 | 80 × 40 | 60 × 30 |
| Passing-go-first sign (square) (cm) | – | – | 80 | 60 |

**Table 5-4** Relationship between minimized font height and operating speeds

| Operating speed (km/h) | Font height (cm) | Operating speed (km/h) | Font height (cm) |
|---|---|---|---|
| $v_{85} \geq 120$ | 60 | $60 \leq v_{85} < 80$ | 30 |
| $100 \leq v_{85} < 120$ | 50 | $40 \leq v_{85} < 60$ | 20 |
| $80 \leq v_{85} < 100$ | 40 | $v_{85} < 40$ | 10 |

principles stipulated in the current *Design Specifications for Highway Traffic Safety Facilities* (JTG D81-2017).

When selecting the types of barriers, the following factors need to be considered: The protective performance of the barriers, the deformation degree of barriers after collision, placement conditions of barriers, the availability of barrier materials, the utilization of existing highway barriers in the region, the life-cycle cost of barriers, the quantities of maintenance work and the convenience level of maintenance, etc.

The performance of the standard sections of highway barrier, transition sections of highway barriers, as well as of the median opening barriers, barriers terminals and crash cushions need to meet the requirements of the *Standards for the Safety Performance Evaluation of the Highway Barriers* (JTG B05-01-2013). The evaluation of barrier design for the highway upgrading and reconstruction project shall also be in compliance with relevant provisions of highway barriers in Section 5.3 of the *Guidelines for Design of Traffic Engineering and Facilities of Expressway Reconstruction* (JTG/T L80-2014).

8 Chapter 3 in the *Guidelines for Design* of *Traffic Engineering and Facilities of Expressway Reconstruction* (JTG/T L80-2014) stipulates the requirements for

evaluation of the current situation and utilization effects of existing traffic control devices and roadside facilities.

The second item, Clause 4.0.8 of the *Guidelines for Design of Traffic Engineering and Facilities of Expressway Reconstruction* (JTG/T L80-2014) stipulates that for road sections with the separation of traffic in the same direction and transition sections with different ways of widening, a scheme for strengthening safety and surveillance facilities shall be provided. A special study could be conducted if necessary. Meanwhile, Clause 7.2.3 (JTG/T L80-2014) provides the requirements for the placement of surveillance facilities. Transition sections with different ways of widening mainly include: the connection among the sections widened on one side and sections widened on both sides of the subgrade sections, the connection among the sections widened on one side of different directions, the connection between directly and separately widened sections. This Clause requires traffic safety and surveillance facilities to be evaluated. In addition, Clause 7.5.8 stipulates that if the night time traffic volume is heavy, the lighting facilities should be installed on the transition sections at the starting point of the same-directional separation section, and on the openings of the belt-like divider for separating same directional traffic flow close to the interchange exit.

9 Speed limit proposals should comply with the regulatory requirements of road safety. After considering and weighing safety and efficiency, a preliminary speed limit scheme could be determined. Because in the detailed design stage the speed limit scheme is a preliminary form, further verification for safety and efficiency is required at the operation stage. Therefore 'should' is used in this *Specifications*. Mainly the types of speed limit and posted speed limit are evaluated.

Investigation indicates that the types of speed limit mainly include spot speed limit, speed zones, speed limit for different vehicle types, speed limit based on both the different lanes and vehicle types, variable speed limits, advisory speed limit, etc. Various types of speed limit have different key points. Since each type of speed limit has its own advantages and disadvantages and scope of application, types of speed limit need to be optimized in accordance with the road factors.

To design the type of speed limit, the following factors ought to be considered, such as laws and regulations, road functions, design criteria, characteristics of vehicle operation, roadside environment and traffic safety characteristics, operability of enforcement etc. Each type of speed limit needs to be matched with posted speed limit values. The selection of posted speed for spot speed limit and speed zones is mainly done by considering the equilibrium features of highway design indicators.

The selection of posted speed or speed limits for different vehicle types is mainly done by considering the speed characteristics of small and large vehicles in China; the selection of posted speed value for speed limits based on both the different lanes and vehicle types is mainly done by considering the characteristics of vehicle speeds and lane functions; the selection of posted speed for variable speed limits is mainly done by considering extreme weather conditions or incidents.

5.4.8 Traffic management proposals need to meet the safety requirements of operating vehicles, construction vehicles, and road builders in highway upgrading and reconstruction projects.

Traffic management proposals should be considered to minimize the traffic impact to the regional road network and to coordinate the construction procedure in traffic management. Clause 4.8.2 of the *Guidelines for Design of Highway Upgrading Reconstruction and Extension* (JTG/T L80-2014) stipulates requirements for the evaluation of traffic management. Chapter 12 presents the requirements for the design of traffic management. The traffic simulation method could be adopted to evaluate technical indicators such as design capacity, service level, and delays in the road network.

# 6 Pre-opening Stage

## 6.1 General

6.1.1 The pre-opening stage provides the opportunity to inspect the post-construction implications of highways and its facilities before being opening to traffic. The safety audit at the stage of pre-opening is a basis for ensuring the safe operation and realizing safe transportation for all road users, and it is a long-term mechanism for ensuring safe traffic and transportation. According to the Law of the People's Republic of China on Work Safety (implemented as of December 1, 2014), production and business units must see to it that safety facilities for their projects to be built, renovated or expanded are designed, constructed, and put into operation and use simultaneously with the principal parts of the projects. Investments into safety facilities must be included in the budgetary estimates of the construction projects.

Clause 13 of Directives of the State Council on Enhancing Road Traffic Safety (Guo Fa [2012] No. 30 Circulation) requires that the development of new roads and road upgrading and reconstruction projects shall be subject to strict safety audit at the time of the as-built (or hand-over)stage. The highway shall not be opened to traffic if highway safety devices does not pass the acceptance inspections. In order to conform with the provisions, this chapter stipulates audit scope, methodology and requirements for pre-opening stage Road Safety Audits.

There are two types of highway acceptance procedure in accordance with the time sequence of project hand-over (or take-over) and as-built in China. The first type is the hand-over and as-built acceptance synchronously after the highway project completion; the second type is conducted after the project completion in the following order: The hand-over stage, trial operation stage, and as-built at end of trial operation stage. The difference between the two types is whether there is a long period (usually more than 2 years) between hand-over and as-built. Usually the road is open to traffic after hand-over procedures.

The requirements in this Chapter are suitable for an RSA at the time of pre-opening for the above two types.

For RSA at the pre-opening stage, it is necessary to collect the detailed design, the documents related to design variations, as well as to inspect and evaluate the quality



H. Wang, *Specifications for Highway Safety Audit*, JTG Standards and Specifications,
https://doi.org/10.1007/978-981-92-3132-4

of the audited project, etc. Auditors should make a thorough on-site investigation by driving, checking safety devices, verifying any potential risk arising from the design variations or changes in roadside surroundings after construction, etc.

6.1.2 The precondition for RSA at this stage is that the major construction engineering and traffic engineering and facilities are completed according to the design documents. Engineering quality is proved from the qualified self-inspection by the construction contractor, the sampling inspection by the supervision contractor and the quality certified by the administration of quality supervision, inspection and quarantine (AQSIQ). In case there is anything related to the verification and evaluation of highway engineering quality during the process of audit, results of corresponding inspection and verification documents may be directly consulted and quoted, if necessary. The auditors may put forward recommendations for additional inspection.

6.4 Audit on Highway Safety Performance

6.4.4 5 Under normal conditions of traffic operation, the lighting, ventilation, fire-fighting and surveillance facilities of tunnels, etc. are relevant to the project quality. In case anything is related to engineering quality inspection and evaluation, the auditors may directly consult and quote the corresponding results of inspection and verification documents, if necessary. The auditors may put forward recommendations for an additional special evaluation. To assess the risk of traffic congestion, fire disasters and other abnormal conditions on traffic safety, if necessary, it is suggested a special evaluation of the actual condition is prepared. Given the complexity of the evaluation and the application feasibility of related provisions, it is advisable to use 'should' in the *Specifications*.

6.4.7 1 To evaluate traffic signs, auditors and drivers are required attend on-site driving, intuitively check the legibility of signs, including reviewing whether the quantity and location of regulatory signs, warning signs, and guide signs are rational; whether the information on the signs is accurate and continuous, and whether the information on signs conforms with and markings, whether roadway geometric design and sign application could be coordinated with the desired driver behaviors, etc.

3 It is necessary to evaluate barriers and median opening barriers according to the actual road surroundings. The auditor ought to verify whether it is necessary to add barriers or adjust the protection level, and assess whether the transition sections of barriers and treatment of barrier terminals complies with the design standards. Once defects of roadside protection are detected, the location, spacing, length, height and other features of designed barriers shall be verified according to the detailed drawings. When it comes to issues related to the engineering quality or structural safety of barriers, conclusions ought to be drawn from the self-inspection by the construction contractor, the sampling inspection by the supervision contractor and the quality certified by the AQSIQ.

4 The anti-glare facilities include anti-glare boards, glare netting and plants. The purpose of placing the anti-glare facilities is to avoid or weaken the glare of light of

vehicles travelling in opposite directions at night. The improper installation of anti-glare facilities may lead to the poor anti-glare implications or poor sight distance. Therefore, the fourth item of this Clause stipulates that the placement of anti-glare facilities and the real effects at night need to be assessed. It should be noticed that the anti-glare facilities installed on the transitional section which connects the separated and integrated subgrades of motorways or Class-1 highways needs to extend full length. The height of anti-glare facilities on median openings of the Class-1 highways should be gradually lowered within a certain distance.

8 The second section of Clause 10.4.1 of *Technical Standards for Highway Engineering* (JTG B01-2014) stipulates that in accordance with the traffic volume, the facilities for surveillance, tolling, communication, power supply and distribution, lighting and maintenance management shall be designed integrally and implemented in phases. Clause 10.4.2 provides for the classification and application scope of the surveillance facilities. Taking into account the requirements for the phased construction of surveillance facilities and the application scope of monitoring facilities of different classes, the evaluation mainly focuses on whether the quantity of monitoring facilities (such as video surveillance, dynamic information release and traffic guidance, etc.) is enough to monitor toll stations, super-large bridges, tunnels, interchanges, intersections, service area, bus stops or the sections etc. Therefore, it is advisable to use 'should' in the *Specification.*

# 7 Operation Stage

7.1 General

7.1.1 The clauses of this Chapter could be applied to the operational highways. The post-evaluation of highway construction is usually carried out after the engineering has been completed, passed as-built acceptance and opened to traffic for over five years in China. *Administration Method for Post-Evaluation of Highway Construction Project* and *Guidelines for Drafting and Writing Method of the Post-Evaluation Report of Highway Construction Project* (Jiao Gui Fa [2011] No. 695 Circulation) requires that RSA reports should be a special report of post-evaluation. According to results of the questionnaire survey in provinces and cities in China, except for the requirements of above circulation, RSAs are still needed for after opening to traffic due to distinct changes of road network operations, traffic volume, roadside conditions, traffic accidents etc. or, at the time of as-built acceptance, major repair or medium rehabilitation, and highway upgrading and reconstruction projects.

7.2 Methodology

7.2.1 The following methods are frequently used to identify the accident-prone sections, including crash counts and crash rates, accumulative frequency curve, quality control, the fuzzy clustering, the safety factor, the expert system, etc. The crash quantities and crash rates are relatively simple and feasible to use in evaluation.

7.3 Overall Audit

7.3.2 Data ought to be collected for the evaluation by survey of traffic flow operation, main causes of traffic crash, accident-prone sections and traffic safety management, etc. The auditor requires to visit managers and technical experts at highway administration, and traffic police or organize dedicated discussions if necessary.

7.3.6 Facilities related to highway EER mainly include the emergency lay-bys, median openings, video surveillance, detection devices of traffic incidents, as well as devices of real-time information and traffic guidance, etc. These facilities directly affect the operating speed of the highway network and efficiency of emergency



H. Wang, *Specifications for Highway Safety Audit*, JTG Standards and Specifications,
https://doi.org/10.1007/978-981-92-3132-4

rescue. Since the promulgation and implementation of the Emergency Response Law of the People's Republic of China in 2007, since crashes do serious harm to society, the highway transportation industry has paid attention to the prevention of, and preparation for, response to emergencies, monitoring and early warning, emergency handling, rescue and relief, and post-emergency rehabilitation and reconstruction, etc. The precautionary plans for responding to emergencies dealing with traffic accidents, adverse weather, tunnel incidents and traffic volume peak hours on major holiday, etc. are gradually incorporated into the daily operation and management of highways, and play a great role in ensuring traffic safety. Therefore, this Clause provides for the evaluation of highway facilities and precautionary plans in response to emergencies. Since the emergency response is a complex system, considering the difficulty of data collection and the feasibility of provisions, it is advisable to use 'may' in the *Specifications*. If necessary, it is recommended to conduct a special evaluation.

7.4 Audit on Highway Safety Performance

7.4.1 3 Observation of vehicle speeds, the sample number, time, location, as well as instrument accuracy and observation error, etc. shall meet the requirements of statistical data analysis. The section speed observation test methods are mainly employed by the video detectors, laser speed detectors, infrared and radar detectors, geomagnetic detectors and other methods.

7.4.2–7.4.11 Although the scope and data of the on-site inspection is stipulated in the Article of the *Specifications*, it can be added and deleted according to the actual condition of the highway project during the evaluation process.

# Technical Terms in Chinese and English

| 序号 | 中 文 词 汇 | 英 文 词 汇 |
|---|---|---|
| 1 | 3s运行速度行程长度 | Distance of 3 seconds at the operating speed |
| A | | |
| 2 | 安全检查清单 | Checklist |
| 3 | 安全交叉停车视距 | Safety intersection stopping sight distance |
| 4 | 安全评估 | Safety evaluation/safety assessment |
| 5 | 暗适应 | Dark adaptation |
| 6 | 凹形竖曲线底部 | The bottom of sag vertical curve |
| B | | |
| 7 | 白洞效应 | The white-hole effect |
| 8 | 避险车道 | Escape ramp |
| 9 | 避险车道引道 | Escape ramp approach |
| 10 | 变速车道 | Speed-change lanes |
| 11 | 标线 | Markings |
| 12 | 标志的支撑方式 | The support of traffic signs |
| 13 | 标志视认效果 | Visibility and legibility of traffic signs |
| 14 | 标志信息 | Information on signs |
| 15 | 不同加宽方式 | Different forms of widening |
| 16 | 部分控制出入 | Partial access control |
| 序号 | 中 文 词 汇 | 英 文 词 汇 |
| C | | |
| 17 | 侧翻事故 | Roll-over crashes |
| 18 | 侧滑和尾随相撞事故 | Skidding and rear-end collision |
| 19 | 超车视距 | Overtaking sight distance |
| 20 | 超高过渡段 | Super-elevation transition |
| 21 | 超高渐变率 | Super-elevation transition rate |

(continued)



H. Wang, *Specifications for Highway Safety Audit*, JTG Standards and Specifications,
https://doi.org/10.1007/978-981-92-3132-4

(continued)

| 序号 | 中文词汇 | 英文词汇 |
|---|---|---|
| 22 | 超高路段 | Super-elevated section |
| 23 | 车道数平衡 | The balance in the number of lanes the equilibrium of lane numbers |
| 24 | 车道转换带 | Lane change-over zones for changing lane |
| 25 | 车行横通道/人行横通道 | Cross passages for vehicles or pedestrians |
| 26 | 冲突点数量 | The number of conflict points |
| 27 | 冲突区域 | Conflict areas |
| 28 | 出口曲线-曲线 | Exit curve-curve |
| 29 | 出口曲线-直线 | Exit curve-straight line |
| 30 | 出入口过渡段 | The entrance/exit transition sections |
| 31 | 次路减速让行 | Give-way on minor road |
| 32 | 次路停车让行 | Stopping on the minor road |
| D | | |
| 33 | 大型构造物 | Large structures |
| 34 | 大中修 | Major repair and medium rehabilitation |
| 35 | 代表车型 | Representative vehicles |
| 36 | 单侧拼宽 | Widened on one side |
| 37 | 单行线标志 | One-way street sign |
| 38 | 当量化平面交叉口个数 | The number of equivalent intersections |
| 39 | 当量化平面交叉口密度 | Equivalent intersection density |
| 40 | 地磁式检测仪 | Geomagnetic detectors |
| 41 | 第三方单位 | The third-party |
| 42 | 动态信息发布 | Dynamic information release |
| 43 | 洞口朝向 | Orientation of tunnel portal |
| 44 | 洞门形式 | Forms of tunnel portal |
| 序号 | 中文词汇 | 英文词汇 |
| F | | |
| 45 | 反光强度等级 | The retro reflectivity levels |
| 46 | 防护设施 | Protection facilities |
| 47 | 防眩板 | Anti-glare panels/glare screen |
| 48 | 防眩设施 | Anti-glare devices |
| 49 | 防眩效果 | Anti-glare performance |
| 50 | 防撞垫 | Crash cushion |
| 51 | 分车道与分车型结合限速 | Speed limit based on both the different lanes and vehicle types |
| 52 | 分车型限速 | Speed limits for different vehicle types |
| 53 | 分流鼻端、合流鼻端 | Nose of diverging areas/nose of merging areas |
| 54 | 分流鼻端通过速度 | The speed passing the gore nose of diverging areas |

(continued)

(continued)

| 序号 | 中文词汇 | 英文词汇 |
|---|---|---|
| 55 | 分区段限速 | Speed zone |
| 56 | 服务区/停车区 | Service area/rest area |
| 57 | 服务水平 | Level of service |
| 58 | 辅助车道 | Auxiliary lane |
| 59 | 辅助设施 | Ancillary facilities |
| 60 | 复合曲线 | Compound curves |
| G | | |
| 61 | 改扩建公路 | A highway upgrading and reconstruction project |
| 62 | 干线公路 | Arterial highway |
| 63 | 刚性固定物 | Fixed rigid objects |
| 64 | 隔离设施 | Facilities for segregation |
| 65 | 工程可行性研究阶段 | Feasibility Stage |
| 66 | 公路安全状况评价 | Audit on highway safety performance |
| 67 | 公路行业相关独立资质 | The qualification of highway industry expertise |
| 68 | 公路现场踏勘 | On-site inspection |
| 69 | 公路项目安全性评价 | Road Safety Audit (RSA), highway safety audit |
| 70 | 公路项目安全性评价规范 (JTG B05-2015) | *Specifications for Highway Safety Audit* (JTG B05-2015) |
| 序号 | 中文词汇 | 英文词汇 |
| 71 | 公路项目安全性评价指南 (JTG/T B05-2004) | *Guidelines for Safety Audit of Highway* (JTG/T B05-2004) |
| 72 | 管理、服务设施 | The administration and service facilities |
| 73 | 光环境变化 | Changes in the light environment |
| 74 | 滚动阻力系数 | Coefficient of rolling resistance |
| H | | |
| 75 | 黑洞效应 | The black-hole effect |
| 76 | 横向摩阻系数 | Lateral friction coefficient |
| 77 | 互通式立体交叉路段 | Interchange sections |
| 78 | 护栏 | Barrier |
| 79 | 护栏标准段 | Standard sections of highway barrier |
| 80 | 护栏的防护性能 | The protective performance of the barriers |
| 81 | 护栏端头 | Barrier terminals |
| 82 | 护栏端头处理方式 | Terminal treatment of barrier |
| 83 | 护栏过渡段 | Transition sections of highway barriers |
| 84 | 护栏最小长度 | Minimum length of barrier |
| 85 | 缓冲设施 | Impact attenuator |
| 86 | 回头曲线 | U curve |
| 87 | 回旋线 | Clothoid |

(continued)

(continued)

| 序号 | 中文词汇 | 英文词汇 |
|---|---|---|
| 88 | 会车视距 | Meeting sight distance |
| 89 | 会车先行标志 | Passing-go-first sign |
| 90 | 混合交通 | Mixed traffic |
| J | | |
| 91 | 积水 | Ponding of water |
| 92 | 激光测速仪 | Laser speed detectors |
| 93 | 急弯陡坡 | Sharp curves and steep grades |
| 94 | 集散一级公路 | Collector-distributor Class-1 highway |
| 95 | 计算净区宽度 | Suggested clear zone width |
| 96 | 加、减速车道长度 | Lengths of acceleration and deceleration lanes |
| 97 | 驾驶模拟舱 | Driving simulator |
| 98 | 驾驶模拟方法 | The method of driving simulation |
| 序号 | 中文词汇 | 英文词汇 |
| 99 | 驾驶期望 | The driving expectation |
| 100 | 驾驶人的工作负荷 | Drivers' workload |
| 101 | 监测与预警 | Monitoring and early warning |
| 102 | 监控设施 | Surveillance and control facilities |
| 103 | 检查站、超限检测站 | Check points |
| 104 | 减速标线 | Deceleration markings |
| 105 | 减速路面 | Speed reduction pavement |
| 106 | 建议限速 | Advisory speed limit |
| 107 | 建筑限界 | Clearance profile |
| 108 | 交工阶段 | Pre-opening Stage/Hand-over Stage/Taking-Over Stage Audit |
| 109 | 交通安全变化趋势 | Trends of traffic safety |
| 110 | 交通安全评价 | Traffic safety audit |
| 111 | 交通安全设施 | Traffic safety facilities |
| 112 | 交通工程及附属设施 | Traffic control devices and ancillary facilities |
| 113 | 交通管理方式 | The method of traffic control/traffic management model |
| 114 | 交通流严重紊流 | Serious turbulence of traffic flow |
| 115 | 交通瓶颈 | Traffic bottlenecks |
| 116 | 交通渠化设施 | Traffic channelization design |
| 117 | 交通事故的空间分布 | The spatial distribution of traffic crashes |
| 118 | 交通延误 | Traffic delays |
| 119 | 交通诱导 | Traffic guidance |
| 120 | 交通运行特点 | Traffic operation characteristics |
| 121 | 交通阻滞 | Traffic congestion |

(continued)

(continued)

| 序号 | 中文词汇 | 英文词汇 |
|---|---|---|
| 122 | 交通组成 | Traffic composition |
| 123 | 交通组织设计 | The design of traffic management |
| 124 | 交织区 | Weaving section |
| 125 | 紧急停车带 | Emergency lay-bys |
| 126 | 禁令标志 | Regulatory signs |
| 127 | 禁止超车路段 | No-overtaking zone |
| 128 | 经验分析法 | Empirical analyses |
| 序号 | 中文词汇 | 英文词汇 |
| 129 | 警告标志 | Warning signs |
| 130 | 竣工图 | As-built drawing |
| 131 | 竣工验收 | The as-built acceptance |
| K | | |
| 132 | 抗滑能力 | Skid resistance |
| 133 | 可变限速 | Variable speed limits |
| 134 | 扩挖隧道 | Expanded tunnel |
| L | | |
| 135 | 雷达检测仪 | Radar detectors |
| 136 | 离心加速度变化率 | The appropriate changing rate of centrifugal acceleration |
| 137 | 立面标记 | Object markers |
| 138 | 连续上坡 | Continuous upgrades/continuous ascending sections |
| 139 | 连续长陡下坡路段 | Continuous long and steep downgrade/descending |
| 140 | 淋水制动 | Braking by dripping water |
| 141 | 路侧冲突因素加权值FRIC | Weighted values of roadside conflict factor, FRIC |
| 142 | 路侧干扰 | Roadside interference |
| 143 | 路侧净区 | Roadside clear zone |
| 144 | 路侧障碍物 | Roadside obstacles |
| 145 | 路面抗滑能力易下降的路段 | Sections prone to reduction in skid resistance |
| 146 | 路线 | Route |
| 147 | 卵形曲线 | Oval curves |
| 148 | 轮廓标 | Delineators |
| M | | |
| 149 | 门架式支撑结构 | Gantry-type support structure |
| 150 | 明适应 | Light adaptation |
| P | | |
| 151 | 爬坡车道 | Climbing lane |
| 152 | 排水不良 | Poor drainage |

(continued)

(continued)

| 序号 | 中文词汇 | 英文词汇 |
|---|---|---|
| 153 | 排水设施 | Drainage facilities |
| 154 | 平安交通 | Safe transportation |
| 序号 | 中文词汇 | 英文词汇 |
| 155 | 平面交叉口密度 | Density of intersections |
| 156 | 平面交叉口折算系数 | Conversion coefficient of intersections |
| 157 | 平坡 | Gentle slope |
| 158 | 平曲线连接形式 | The form of connections of horizontal curves |
| 159 | 平曲线路段 | Sections of horizontal curves |
| 160 | 平直路段 | Tangent sections |
| 161 | 平纵指标较低路段 | Road sections with constrained horizontal and vertical alignment features |
| 162 | 评价标准 | Evaluation criterion |
| 163 | 评价程序 | Audit procedure/audit process |
| 164 | 评价方法 | Methodology |
| 165 | 评价阶段 | Stages of audit |
| 166 | 评价结论 | Findings and recommendations/findings and suggestions |
| 167 | 评价内容 | Scope of audit |
| Q | | |
| 168 | 期望速度 | Desired speed |
| 169 | 起讫点 | Origin-destination |
| 170 | 气象灾害多发路段 | Road sections susceptible to weather hazards |
| 171 | 桥梁护栏与路基护栏 | Bridge railing and grade barrier |
| 172 | 桥面泄水孔 | Discharge holes on bridge decks |
| 173 | 桥隧相连路段 | Bridge-tunnel sections |
| 174 | 桥头急弯路段 | Bridges preceded by sharp curve |
| 175 | 渠化设计 | Channelization design |
| 176 | 曲线中点 | Mid-point of the horizontal curve |
| 177 | 全部控制出入 | Full access control |
| 178 | 全段统一限速 | Single speed limit on the all road sections |
| 179 | 全寿命周期成本 | The life-cycle cost |
| 180 | 人行道 | Sidewalk |
| 181 | 入口曲线-曲线 | Entrance curve-curve |
| 182 | 入口直线-曲线 | Entrance tangent-curve |
| 序号 | 中文词汇 | 英文词汇 |
| S | | |
| 183 | 闪光警告信号灯 | Flash warning signal lights (Operation Yellow Change and Red Clearance Intervals) |
| 184 | 上跨桥梁 | Overpass bridge/bridges crossing |

(continued)

(continued)

| 序号 | 中文词汇 | 英文词汇 |
|---|---|---|
| 185 | 设计变更 | Design variations |
| 186 | 设计审查 | The approval of detailed design |
| 187 | 设计要素评价 | Audit of design elements |
| 188 | 设计一致性 | Design consistency |
| 189 | 失控车辆 | Uncontrolled vehicles/an errant vehicle/out-of-control vehicle |
| 190 | 施工期间 | Construction period |
| 191 | 施工图设计阶段 | Detailed design stage |
| 192 | 识别视距 | Decision sight distance |
| 193 | 实地驾驶 | On-site driving |
| 194 | 事故车型 | Vehicle type involved in the crashes |
| 195 | 事故率 | Crash rate |
| 196 | 事故频发路段 | Crash-prone sections |
| 197 | 视线诱导设施 | Visual guiding device |
| 198 | 收费方式 | Method of toll collection |
| 199 | 竖曲线 | Vertical curve |
| 200 | 速度控制和交通管理措施 | Measures of speed control and traffic management |
| 201 | 速度控制设施 | Speed management facilities |
| 202 | 隧道洞口 | Tunnel portals |
| 203 | 隧道路段 | Tunnel sections |
| 204 | 隧道群 | Tunnel-chains |
| 205 | 隧道与互通式立体交叉相连路段 | Tunnel-interchange sections |
| 206 | 隧道照明 | Tunnel lighting |
| T | | |
| 207 | 停车视距 | Stopping sight distance |
| 序号 | 中文词汇 | 英文词汇 |
| 208 | 通视三角区 | Sight triangle |
| 209 | 同深度走廊带比选方案 | Alternative corridor proposals with the same level of details |
| 210 | 同向车道分隔带 | The belt-like divider for separating the same direction traffic flow |
| 211 | 同向分离路段起点的过渡段 | The start of separation of traffic in the same direction |
| 212 | 同一路段 | The same segment |
| 213 | 同一路段运行速度与设计速度协调性 | Speed consistency between operating speed and design speed on the same segment |
| 214 | 突起路标 | The raised pavement markers |
| 215 | 土建工程 | Civil works |

(continued)

(continued)

| 序号 | 中 文 词 汇 | 英 文 词 汇 |
|---|---|---|
| 216 | 推荐方案 | Recommended proposals |
| W | | |
| 217 | 弯坡组合路段 | Grade-curve sections |
| 218 | 危险路段 | Dangerous segments |
| 219 | 无优先交叉 | No-priority |
| X | | |
| 220 | 行车安全评价 | Driving safety audit |
| 221 | 行车道 | Traveled-way/carriageway |
| 222 | 行驶轨迹 | Driving trajectory |
| 223 | 限速方案 | Speed limit proposals |
| 224 | 线形诱导标 | Alignment guiding device |
| 225 | 线形指标 | Alignment features |
| 226 | 相交公路等级 | Classification of intersected highways |
| 227 | 相邻出入口间距 | Spacing between successive exits/adjacent entrances or exits on a route or ramps |
| 228 | 相邻路段 | Successive geometric feature segments |
| 229 | 相邻路段运行速度协调性 | Operating speed consistency between successive geometric feature segments |
| 230 | 泄水能力 | The drainage capacity |
| 序号 | 中 文 词 汇 | 英 文 词 汇 |
| Y | | |
| 231 | 烟雾设计浓度 | Fog design concentrations |
| 232 | 养护维修作业控制区 | Work zone |
| 233 | 业主单位 | Owner's agency |
| 234 | 一阶段施工图设计 | Single stage detailed design |
| 235 | 应急救援和紧急疏散 | Evacuation emergency and rescue (EER) |
| 236 | 应急预案 | Emergency response plan |
| 237 | 预测交通量 | Forecast traffic volumes |
| 238 | 运行速度 | Operating speed |
| 239 | 运行速度差值 | Operating speed differentials |
| 240 | 运行速度梯度 | Operating speed gradient |
| 241 | 运行速度突变 | The sudden changes in operating speed |
| 242 | 运行速度协调性 | Operating speed-based design consistency |
| 243 | 运行速度预测模型 | Operating speed prediction model |
| 244 | 运行速度折减值 | Reduction value of the operating speeds |
| 245 | 运营阶段, 后评价 | Operating stage |
| Z | | |
| 246 | 匝道出、入口 | Ramp entrances and exits |

(continued)

(continued)

| 序号 | 中文词汇 | 英文词汇 |
|---|---|---|
| 247 | 匝道出入口线形 | The geometric alignment of on-ramps and off-ramps |
| 248 | 匝道基本路段 | Ramp typical sections |
| 249 | 匝道控制曲线 | Alignment control curve of a ramp |
| 250 | 长下坡接小半径曲线路段 | Long downgrades preceding small-radius curves |
| 251 | 长效机制 | Long-term mechanism |
| 252 | 直接式减速车道 | Direct deceleration lane |
| 253 | 指路标志 | Guide sign |
| 254 | 指示标志 | Indicative signs |
| 255 | 制动距离 | Stopping distance |
| 256 | 制动坡床材料 | Materials of arrester bed |
| 257 | 制动砂床 | Arrester bed |
| 258 | 制动失效 | Failed brakes |
| 259 | 中央分隔带护栏的防撞等级 | The protection levels of median barrier |
| 序号 | 中文词汇 | 英文词汇 |
| 260 | 中央分隔带护栏的形式 | Types of median barrier |
| 261 | 中央分隔带开口护栏 | Median opening barrier |
| 262 | 重大、特大交通事故 | Fatal and serious traffic crashes |
| 263 | 主路优先 | Main road priority |
| 264 | 主线 | Mainline/main route |
| 265 | 主线的相邻出口或入口 | Adjacent exits or entrances of main route |
| 266 | 主线分、合流 | Diverges and merges of the mainline |
| 267 | 主线收费站 | Toll station/plaza on main route |
| 268 | 转弯附加车道 | Auxiliary turning lanes |
| 289 | 资质证书 | Qualification certificates |
| 270 | 自然灾害 | Natural disaster |
| 271 | 自由流状态 | Free-flow conditions |
| 272 | 总体评价 | Overall audit |
| 273 | 纵坡路段 | Sections on longitudinal grade |
| 274 | 纵坡路段运行速度折算模型 | The conversion model of longitudinal grade sections |
| 275 | 走廊带方案 | Corridor proposals |
| 276 | 阻拦网 | Arresting nets |
| 277 | 最低限速值 | Minimum speed limit value |
| 278 | 最低运行速度 | Minimum operating speed |
| 279 | 最小净距 | The minimum net distance |

MIX
Papier aus verantwortungsvollen Quellen
Paper from responsible sources
FSC® C105338

If you have any concerns about our products,
you can contact us on
**ProductSafety@springernature.com**

In case Publisher is established outside the EU,
the EU authorized representative is:
**Springer Nature Customer Service Center GmbH**
**Europaplatz 3, 69115 Heidelberg, Germany**

Printed by Libri Plureos GmbH
in Hamburg, Germany